HUMAN
GENETIC
ENGINEERING

For more by and about Pete Shanks,
including web links for further
information, see his website
www.wordsontheweb.com.

HUMAN
GENETIC
ENGINEERING

A GUIDE FOR ACTIVISTS, SKEPTICS,
AND THE VERY PERPLEXED

■ PETE SHANKS ■

NATION BOOKS
NEW YORK

HUMAN GENETIC ENGINEERING
A Guide for Activists, Skeptics, and the Very Perplexed

Copyright © 2005 by Pete Shanks

Published by
Nation Books
An Imprint of Avalon Publishing Group
245 West 17th St., 11th Floor
New York, NY 10011

AVALON
publishing group incorporated

Nation Books is a co-publishing venture of the Nation Institute and
Avalon Publishing Group Incorporated.

Library of Congress Cataloging-in-Publication Data is available.

ISBN 1-56025-695-8

9 8 7 6 5 4 3 2 1

Book design by Pauline Neuwirth, Neuwirth and Associates, Inc.
Printed in the United States of America
Distributed by Publishers Group West

CONTENTS

1

CROPS AND CATTLE AND FISH AND ... PEOPLE?

■

■

■ INTRODUCTION ■

THE PUBLIC DEBATE over human biotechnology is just beginning. Experts have been discussing the implications of genetic research for decades, but only in the last few years have these topics begun to reach the political agenda.

Cloning has hit the headlines, and so has the partly related question of embryonic stem cell research. We still don't have a federal law to regulate either of them, and research continues anyway. Meanwhile, some practitioners in the fertility industry are looking to expand the market for their services to include not only sex selection but "designer baby" options. Once again, the political questions remain: Who decides what is acceptable and how should those decisions be enforced?

These developments are occurring against a background of people forgetting—or misrepresenting—the terrible history of eugenics in the twentieth century. Eugenics was not primarily developed by the Nazis, although they used it to justify their prejudices. Eugenics was an appalling idea mostly advocated by well-meaning people. It's coming back, this time as a consumer option, in a high-tech form that has appropriately been dubbed "techno-eugenics."[1]

The US has no effective and comprehensive laws governing the use and abuse of these technologies. Scientists as well as the general public would benefit from a system that both provided oversight against abuse and protected the legitimate rights of researchers. Until we can develop something that provides those reassurances, this debate is only going to get more urgent.

Genetically engineered (GE; see **Box 1.1**) food is routinely sold in American supermarkets. GE mice are created and cloned all the time for scientific experiments. Cloned cattle are being created and auctioned off. Glowing GE fish are being sold as pets, and a company called Genetic Savings and Clone is even selling cloned cats. Every month, it seems, there's another "breakthrough." What's next?

1.1

ACRONYMS, EUPHEMISMS, AND INSULTS

GE   Genetic Engineering or Genetically Engineered
GM   Genetically Modified
GMO   Genetically Modified Organism

This book uses "GE" as both noun and adjective. In practice, all genetically engineered plants and animals are genetically modified; the terms are effectively synonymous.

Scientifically altered is sometimes used to avoid saying "GE" or "GM."

Somatic Cell Nuclear Transfer (SCNT) means "cloning" and is often used by those in favor of embryonic research, especially those whose only concern about cloning is that it doesn't work safely enough to use on people (see **Chapter 3**).

Designer babies is a term that annoys some advocates of Human GE, but has taken hold as useful shorthand for GE babies in nontechnical settings; some people see nothing wrong with either the expression or the idea.

Frankenfoods is a flat-out insult, but undeniably clever.[2]

■ GENETICALLY ENGINEERED PEOPLE? ■

THE PROSPECT OF Human GE sends shivers down most spines. On balance, we don't like genetically modified food; we dislike even more the idea of tinkering with animals; and, according to a 2003 survey conducted for the Pew Initiative on

Food and Biotechnology:[3] "Consumers are least comfortable with genetic modifications of humans." (In the report, prepared by The Mellman Group, Inc., "consumers" seems to be used as a synonym for "Americans.")

Whether or not we are uncomfortable, Human GE is an issue we have to face and deal with immediately. "Gene therapy" has been in trials for years, so far with little success (see **Chapter 6**); if it works, it does, in a sense, create GE people, since new genes are introduced to supplant the patient's original ones. Cloning is thus far hoax and hype, but could be a real problem in the near future (see **Chapter 3**). Sports authorities such as the World Anti-Doping Association (WADA) are taking the imminent prospect of "gene doping" very seriously (see **Chapter 7**). And that's just the start.

Not all genetic technologies are necessarily bad when applied to humans. Genetic research may lead to targeted drugs; that could mean more effective (though probably expensive) treatments for individuals.[4] But at the other extreme, the deliberate creation of semihuman slaves would be a nightmare. Somewhere in between are genetic testing of adults (popular, but it could lead to discrimination) and cloning of babies (very unpopular, but a few people actually want it)—not to mention the testing and selection, and possibly even active engineering, of embryos.

Therefore, one way to look at the complicated web of issues related to Human GE is to place them on a continuum running from what we all (or almost all) agree is good to what we all (or almost all) agree is bad (**Figure 1.1**).

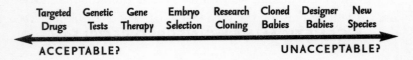

| Targeted Drugs | Genetic Tests | Gene Therapy | Embryo Selection | Research Cloning | Cloned Babies | Designer Babies | New Species |

ACCEPTABLE?   UNACCEPTABLE?

Figure 1.1 Where do we draw the line?

People disagree about what goes exactly where on the continuum. Some disability activists, for example, are even more opposed to embryo selection than to research cloning, since they are particularly concerned about increasing discrimination against the disabled (see **Chapter 11**). We should remember, too, that a compromise position is not necessarily correct; and that some kind of three-dimensional web would be a more realistic representation. In general, however, the question is obvious: *Where do we draw the line?*

■ OUR LEGACY ■

Human GE is sometimes presented as if it were a simple extension of cosmetic surgery. If you can get your nose changed, or your wrinkles smoothed out, why shouldn't you have a technician "fix" whatever gene it is that caused the undesirable effect?

You probably can't; or at least we are a long way from being able to predict accurately what will happen if we fiddle with our genes. But even if we could, if you were to change *your* genes and then have a child, you would pass on the changes as part of your genetic legacy (see **Chapter 2** for more details). Which leads us to perhaps the most important single question Human GE raises: *Should we—if we can—change our children's genes?*

If your immediate response is, "Designer babies? No thanks," you're in good company. Most people agree (see **Chapter 9**). The technical term for changing genes that then get passed on to offspring is **Inheritable Genetic Modification (IGM)**, which has been described, in *Nature*, as "Biology's Last Taboo."[5]

But there are people who want to break the taboo, and they are actively promoting the idea (see **Chapter 10**). James Watson, the codiscoverer of the structure of DNA, is probably the most famous, while Lee Silver and Gregory Stock are academics who have become leading popularizers of the concept. Advocates note, accurately, that many of us already in a negative sense "select" genes we *don't* want our children to have, by testing for various crippling genetic diseases and deciding not to carry the afflicted to term. Some of them go on to argue that "choosing" genes we don't have and "giving" them to our children is not really different—*even if the genes we choose never existed in people before.*

Some of the "transhumanists," "extropians," and the like propounding these theories are, frankly, kooky; but a surprising number of enthusiasts for changing our species, or even creating a new one, work at first-rate universities such as Princeton, UCLA, Stanford, and many others. They include winners of Nobel prizes, heads of research institutes, and other distinguished scientists with regular access to mainstream media.

What are they trying to do? "Improve" us. Consider one specific example: Jay Neitz, a professor of cell biology at the Medical College of Wisconsin, is working on—"with" seems inappropriate—monkeys that have limited color vision because they only have two cones in the eye, compared with our three. His idea is to engineer in a third cone, using the techniques of gene therapy. He then hopes to use the procedure not only to cure colorblindness in humans but to offer us a *fourth* cone, possibly one tuned to infrared.[6] He doesn't know whether our brains can process the information from a fourth cone, but believes that they could.

In the abstract, that might be interesting science. In the particular, just trying to do the experiment in a human would be grossly abusive. (For further discussion, see **Chapter 3**; some would say the research is unethical in animals too.) And who decides it's a legitimate field of inquiry? The individual researcher? Human and animal rights activists? Or society as a whole? And how do we make those decisions and enforce them?

Promoters of Human GE have been trying for several years to persuade the public that Human GE is not only acceptable but desirable, if not inevitable. So far, the public has remained largely unconvinced, but the advocates of Human GE show no sign of giving up.

■ OVERSIMPLIFICATION AND RHETORIC ■

UNFORTUNATELY, MANY EXPERTS oversimplify the questions for public consumption. The "debate" is all too often presented as though human genetic engineering is (pick one)

▶ the greatest medical breakthrough in history
▶ the most terrible prospect since Eve ate the apple

Worse yet, the second alternative is often put forward by supporters of GE as a "straw man" to be demolished. This is not helpful. Nor is the—usually politely disguised, but not always—name-calling that has gone along with the debate: Opponents are Luddites![7] Supporters are Fascists![8]

People don't necessarily fall into simple categories. Human GE is often portrayed as a battle between science and religion, but that's a distortion at best. To name but one prominently religious scientist: Francis Collins, the director of the Human Genome Project for over a decade, is a committed and public Christian. He is strongly in favor of research, obviously, but has reservations about how it should be applied. "My own position is to be intensely conflicted," he has admitted.[9] That's inevitable, and common. Protestant theologians take different positions on different aspects of Human GE, and lay Catholics do not necessarily agree with the Pope.[10] *All* of us span categories and defy easy definition.

Some scientists are certainly optimistic, but others are extremely *pessimistic* about the near-term medical promise of high-tech genetic research; they talk in terms of "fifty years" and even of permanent theoretical obstacles.[11] In public, certainly, and probably in the field as a whole, the real skeptics are a minority, but even committed advocates of research, such as Stanford's Professor Irving Weissman, are more cautious than some patients might like. "We are mindful that this field [embryonic stem cells, in particular] has been overhyped," he said in February 2004, but when it comes to lowering expectations, "I feel like I'm shouting into the wind."[12]

■ LEAVE IT TO THE EXPERTS? ■

MANY EXPERTS ALSO claim a kind of privilege: We are scientists and therefore know what should be done. *Or:* We are bioethicists and therefore . . . *Or:* We are theologians and therefore . . . *Or just:* We are much smarter than you are, and therefore . . .

Nonsense. The issues around Human GE are ones that we all can and should know about—as citizens (we need some new laws), as consumers (we're going to have to make personal choices), and especially as people: We need to come to an agreement, as a society, both national and global, about what is and is not acceptable. Sounds hard? Really, it shouldn't be. That's the message of this book.

■ POLITICAL DECISIONS ■

YOU DON'T NEED to be a nuclear physicist to have an opinion on nuclear energy. You *do* need to be a qualified scientist to conduct a risk assessment; but deciding whether the risk is worth running, that's political. That's a decision that belongs with all of us.

The same is true of Human GE: We—the general public—need to know what the risks and possibilities are; and then we need to decide where to draw the lines. What is acceptable? What is not? How are the boundary issues, of which there will be more and more as science progresses, going to be decided?

There are some regulations about Human GE in the US, but the system is completely inadequate (see **Chapter 12**). Mostly it relies on the ethical codes of the various professional organizations, and on the federal government's ability to attach strings to federal funding of research. That leaves far too many loopholes for the unscrupulous. Other countries, including Britain and Canada, have put into place comprehensive systems of oversight; the US needs one badly.

■ DIFFERENT PERSPECTIVES ■

PEOPLE BRING MANY different perspectives to the discussion about Human GE, each of which provides a lens through which the issues can be viewed (see **Chapters 10** and **11**). Most of us routinely use several of them, but may benefit from being reminded of others. Here's an incomplete, alphabetical list:

- disability rights
- environmentalism
- feminism
- health equity
- indigenous rights
- libertarianism

- patients' rights
- race
- religion
- reproductive rights
- science
- sexual orientation

Some viewpoints may seem incompatible: The rights of the *community* may conflict with the rights of the *individual*; or the rights of *researchers* and those who benefit from their work may conflict with the rights of animal or even human *subjects*. Sometimes just bringing these conflicts into the open helps to resolve them; sometimes they cannot easily be resolved.

We do, however, need to start by acknowledging that all of them have legitimacy, and we need to figure out how to discuss the issues with people coming from each angle. Citing the *Journal of Molecular Biology* won't necessarily convert a Catholic; citing *Das Kapital* isn't guaranteed to convince a Darwinist. Those who are absolutely certain of their position—be they conservative Christians, radical enviros, research scientists, or those coming from any other definable perspective—still should take the time to understand and communicate with the competition. It's the only way to persuade them!

That's how politics is *supposed* to work. But it often doesn't.

▪ ABORTION POLITICS ▪

Some of the strongest supporters of a woman's right to choose are also vehement opponents of the misuse of Human GE. Some of the most vehement opponents of the misuse of Human GE are also among the strongest opponents of abortion. This is confusing.

Proponents of Human GE, quite deliberately, try to attract support by depicting genetic modification as a "choice" (see **Chapter 5** for a discussion of marketing so-called "enhancements"). This is a political tactic, and a rather effective one. It appeals to their core constituency of affluent, centrist technophiles (see **Chapter 9**), while also reaching out to progressives and aiming to keep them separated from "anti-choice" conservatives. The message, of course, is: Those anti-abortion people are our common enemy. The retort is: Human GE is something that *uses* women, views them primarily as creatures that make babies, and values them largely, if not only, in terms of their offspring.

See how easy it is to raise the temperature and confuse the issue?

It is true that people who are opposed to abortion on principle are generally opposed to the genetic manipulation of embryos, for religious reasons. Not all are, however; and the reverse is certainly not true. Many pro-choice progressives are deeply concerned about biotechnology, for a variety of reasons, which have little or nothing to do with abortion (see **Chapter 11**).

In practice, some questions do seem to arise in both the abortion debate and the Human GE one, and they can lead to contradictory answers. For example, some feminists are concerned that any restrictions on embryo research will presage a rollback of abortion rights; others that failure to restrict such research will lead to massive exploitation of women (see **Chapter 4**). And that discussion is simply within the pro-choice feminist camp—when you broaden it to include members of the anti-abortion lobby, things do get complicated.

Complicating the issue works, on balance, to the benefit of those who favor Human GE, because at present we have a political stalemate around regulation of the technologies. To move beyond it, we have to recognize that, actually, the whole concept of "enemies" is unhelpful. Some self-identified conservatives are strong supporters of biotech in all its forms, but that doesn't mean progressives should be against it; some conservatives are strong opponents, which doesn't mean progressives should be in favor of it. There may in fact be a lot to be gained by forging temporary, issue-based agreements with those whom we often oppose.

Disentangling the issues of Human GE from questions of abortion politics is hard, given recent history, but necessary. GE is not a simple debate, it is a multifaceted discussion. Any attempt to force it into a prefabricated mold distorts and confuses the issue.

■ INEVITABLE? NO! ■

PEOPLE AT THE extreme end of the "posthuman" spectrum, especially, like to claim that Human GE is bound to happen; Gregory Stock's original subtitle for his book *Redesigning Humans* was *Our Inevitable Genetic Future*.[13] Anything we can invent will be used, they say, so we might as well accept it. Besides, we have no right to stop people doing what they want to do, especially with their own children, so if some people want GE children, they're bound to have them.

This is a gross overstatement that bears little relation to the world in which we actually live. Society has made and continues to make many decisions about what is acceptable behavior, including outright bans on:

▶ incest
▶ child abuse
▶ child labor
▶ slavery
▶ and many other cruelties

Incest still occurs; so does child abuse—including whippings and beatings that were, until recently, thought to be acceptable—and child labor, although not so much in the US as in the foreign sweatshops that make much of the clothing we buy. (And slavery was, not so long ago, part of the fabric of American life.) We have chosen to try to safeguard children from all these horrors, and have at least made them much less common than they used to be, even though significant numbers of abusive parents still perpetuate them.

Is the use and abuse of technology any different? Not really. Few of us are as selective as the Amish in adopting modern inventions, or want to be, but there are global—or nearly global—bans on, for example:

▶ nuclear testing
▶ torture of prisoners of war
▶ chemical weapons
▶ landmines

The US has a startlingly poor record on these issues, but certainly without the treaties in place there would be far more nuclear proliferation, much worse abuses of POWs, and many more stocks of, and doubtless uses of, chemical weapons. The landmines effort is relatively new, but also promises to reduce a scourge of civilian populations in war torn regions.

These efforts were not 100 percent successful; but they helped. There is plenty of precedent for taking national and global action to prevent, reduce the spread of, or simply regulate dangerous technologies. There are already initiatives at the United Nations to develop international treaties regulating Human GE (see **Chapter 12**). Its uncontrolled use is certainly not inevitable.

▪ CRYING WOLF, DELIBERATELY ▪

"THE CLONES ARE coming, the clones are coming!"[14] The mass media had a field day in the last week of 2002 and the beginning of 2003 with the announcement by a space alien cult that they had cloned a baby. (They've never exactly admitted that it was a hoax, but they've never produced the baby either; see **Chapter 3**.) This followed a furor two years earlier when a couple of mavericks announced that they were going to clone people. (As of early 2005, neither of them has admitted failure either.) Much earlier, in 1978, a journalist then considered respectable produced a whole book describing the supposed creation of a clone; a scientist he cited sued him and, when he failed to produce evidence, the court ruled that the book was "a fraud and a hoax."[15]

All the respectable authorities weighed in to condemn these attempts, including the Biotechnology Industry Organization (BIO) and professional associations like the American Society for Reproductive Medicine (ASRM).[16] They were shocked, shocked to discover that anyone would try something so unproven, so dangerous and so unlikely to succeed.

Some of them, however, carefully avoided condemning cloning outright. The National Academy of Sciences called for a ban on safety grounds, but:[17]

The study panel did not address the issue of whether human reproductive cloning, even if it were found to be medically safe, would be—or would not be—acceptable to individuals or society.

Some, like Nobel laureate Paul Berg, even explicitly said that once the bugs were worked out, they would approve of cloning (see **Chapter 10**). And the real boosters of Human GE mostly kept their heads down. They believe we'll get used to the idea. In fact, they're counting on it.

The concept of "thinking about the unthinkable" has become commonplace since 1962, when Herman Kahn published a book of that name about nuclear war. It's a neat phrase, but meant to be self-canceling. The more you think about something, the more plausible it seems, the more inevitable it becomes. This is a serious risk with Human GE, and may partly explain why so many professional bioethicists become "yes, but" boosters of the technology.

Fortunately, however, the general public does not seem convinced. Polls show as much opposition to cloning as ever—90 percent, give or take. Informed discussion does not have to lead to acceptance; thinking about these issues is in fact the only way to *stop* regrettable outcomes.

■ WHAT THIS BOOK DOES ■

THIS IS A book about Human GE. It does *not* cover the important and related topics of Plant and Animal GE. There are several excellent books about GE Food, listed in the **Appendix**; not nearly enough has been published on GE Animals, but many of the food activists combine the issues.

No single volume could possibly present all the research and all the arguments about Human GE. This book presents an overview, with pointers to where you can learn more. Each chapter includes lists of free documents available on the web, as well as printed materials available from the library or bookstore. These are also listed on the author's website, at www.wordsonthe web.com/HGE, where other materials are posted, including informational leaflets, suitable for printing, copying and distributing. There is a comprehensive list, including contact details for activist groups, in the Appendix.

Between the book itself, the references, and the links, you'll be well equipped to learn enough to make smart decisions about our future as humans. Pass it on!

■ FURTHER READING ■

Free Documents from the Web

The New Technologies of Human Genetic Modification: A Threshold Challenge for Humanity by the Center for Genetics and Society (CGS) in Oakland, California, includes essays and key quotes from both advocates and opponents of the new human genetic technologies. A 56-page pdf file (565k) is at www.genetics-and-society.org/resources/cgs/newtechs.pdf.

"The Quiet Campaign for Genetically Engineered Humans," a 5-page article by CGS Director Richard Hayes, first published in *Earth Island Journal*, is at www.genetics-and-society.org/resources/cgs/2001_earthisland_hayes.html.

"Yuppie Eugenics: Creating a world with genetic haves and have-nots," by Ruth Hubbard and Stuart Newman, two professors and founding members of the Center for Responsible Genetics (CRG) in Cambridge, Massachusetts, is at www.zmag.org/ZMag/articles/march02hubbard-newman.htm.

Why Should I Be Concerned About Human Genetics? is a 12-page pdf (400k) published by Human Genetics Alert in the UK, in A4 format, so Americans should check the "shrink oversized pages to paper size" option when printing; www.hgalert.org/briefings/briefing1.PDF.

WorldWatch, July/August 2002, "Beyond Cloning" Special Issue is available as two pdf files, 10 and 28 pages (288k, 1.1mb), including 16 different articles, from www.worldwatch.org/pubs/mag/2002/154.

Books

Bill McKibben, *Enough,* Henry Holt and Company, 2003; a superb narrative overview from a noted environmentalist

Brian Tokar, ed., *Redesigning Life: The Worldwide Challenge to Genetic Engineering*, Zed Books, 2001; an important collection of essays
Francis Fukuyama, *Our Posthuman Future: Consequences of the Biotechnology Revolution*, Farrar, Straus and Giroux, 2002
Jeremy Rifkin, *The Biotech Century*, Tarcher/Putnam, 1998
Andrew Kimbrell, *The Human Body Shop*, Regnery Publishing, 1997

▪ ENDNOTES ▪

1 The term "techno-eugenics" was coined in 1999 by Richard Hayes, who later cofounded the Center for Genetics and Society; see Box 11.1.

2 "Frankenfood" first appeared in print in a letter to the *New York Times* by Paul Lewis, a Boston professor of English, in June 1992.

3 www.pewagbiotech.org/research/2003update/

4 Individualized medicine could not only be expensive for patients but also take the most profitable products away from pharmaceutical companies. John D. Rhodes, "Beyond the Blockbuster," *Bio-It World*, 08/13/03

5 *Nature*, 413:12–15, 09/06/01

6 *Slate*, 03/05/03

7 *San Diego Union-Tribune*, 06/22/01

8 *Biodemocracy News* 43, 08/03: "[O]rganic food is not going to taste that good in a fascist state." Not a bad line, but is it persuasive?

9 Jordan Lite, "Where's God in the Machine?" *Wired*, 07/24/00

10 The Pew Forum on Religion and Public Life, *Cloning Adam's Rib: A Primer on Religious Responses to Cloning*, 03/02

11 Brian Alexander, *Rapture: How Biotech Became the New Religion*, Basic Books, New York, 2003, p. 204; Barry Commoner, "Unraveling the DNA Myth," *Harper's*, 02/02

12 *New York Times*, 02/15/04

13 Gregory Stock, *Redesigning Humans: Our Inevitable Genetic Future*. Houghton Mifflin: New York, 2002. The 2003 paperback edition, published by Mariner Books, has a different subtitle: *Choosing Our Genes, Changing Our Future*.

14 Dave Koehler at Phillyburbs.com

15 *Washington Post*, 12/31/02; David Rorvik, *In His Image: The Cloning of Man*, New York, J. B. Lippincott Co., 1978

16 *Washington Post*, 01/27/01; *Christian Science Monitor*, 02/07/01

17 "U.S. Policy-makers Should Ban Human Reproductive Cloning," National Academy of Sciences, 01/18/02, press release announcing their report *Scientific and Medical Aspects of Human Reproductive Cloning*

2

THE DEVELOPING SCIENCE OF GENETICS

■ INTRODUCTION ■

I F YOU DON'T know how genes work, you're in good company. No one does. Not exactly. It almost seems that the more we learn, the less we really know.

That doesn't mean we know nothing about the process, and it certainly doesn't mean we can't affect it. We can—and do—swap genes in and out of completely unrelated species. We can:

- create bunnies that glow in the dark[1]
- make lab mice with specified genes turned permanently on or off
- force salmon to give birth to trout[2]
- produce crops that are resistant to one particular weed-killer

There are problems with all these applications. Most are inefficient, and at least some have unexpected side effects or ramifications. They do essentially work, at least some of the time, but there are an awful lot of surprises—most of them unpleasant—along the way.

Partly, that's just life near the cutting edge of science. Physics and cosmology have massive unanswered questions, and they don't stop us sending rockets to Mars. But when those missions fail, as they rather frequently do, the problems are usually mechanical, rather than theoretical or acutely surprising.[3] Biology is still in the scientifically fascinating place where experimental failures point the way to greater theoretical, not just technological, understanding.

We have not been able, despite fifteen years of intensive effort, to perform "gene therapy" successfully on people (see **Chapter 6**), let alone "enhancement" (see **Chapter 7**). Not only are there ethical questions about human experiments (see also **Chapter 3**, on cloning), and social questions of exploitation and abuse (see **Chapter 4**, among others)—there also remain what may be fundamental scientific questions.

This chapter can only give the sketchiest of overviews; entire textbooks are devoted to the subjects involved. Most of the discussion in the rest of the book does *not* require that you master the technical details, even at the very simple level presented here. Given the general idea of, say, cloning, it's perfectly possible to discuss it sensibly without knowing much about how it is done.

Scattered through the text are questions and observations that give some idea of how much we all still have to learn. For example, the deceptively simple question, how many genes do we have? (See **Box 2.1** and, on the lighter side, **Box 2.2**.)

2.1

HOW MANY GENES DO WE HAVE?

PEOPLE PROBABLY have 20–25,000 genes. That's according to an October 2004 paper in the authoritative journal *Nature*, but the conventional wisdom does keep changing, and it's worth noting that ±10 percent is a fairly healthy margin for error.[4] We certainly have nowhere near as many as most experts thought as recently as 2000 (see also **Box 2.2**).

A few computational biologists have made estimates below 20,000, and many scientists are reluctant to abandon a prediction of around 40,000; a few go as high as 70,000. Dr. Craig Venter, who headed one of the Human Genome consortiums, estimated 26,000 in 2001 but cagily suggested that:[5] "It will probably take 5 or 10 years to have a really accurate count plus or minus 100 genes."

As a matter of historical curiosity, until the 1990s, most textbooks guessed that we had (about) 100,000 genes "for no other reason than it was a convenient round number."[6] That began to change with a 1994

2.1 article in *Nature Genetics*, which estimated 60–70,000 and prompted this furious reaction from one genetic entrepreneur:[7] "What the hell do you think you're doing, saying there are only 60,000 genes? I just sold 100,000 genes to SmithKline Beecham!"

The real answer is that the precise number is an interesting but fundamentally unimportant piece of trivia. Here's another that puts it into perspective: If the 2004 best guesses are even close to correct, then *rice* has twice as many genes as people do. Basmati rice, the *indica* variety, has 46–56,000 genes.[8]

2.2 ## THE GREAT GENE SWEEPSTAKES

IN THE summer of 2000, when the first draft of the human genome was announced (see **Box 2.5**), geneticists were seriously debating how many genes people have, because they thought it would soon be settled. At the annual meeting of experts in Cold Spring Harbor, New York—actually, one evening in a bar—Dr. Ewan Birney hit upon the idea of having a sweepstakes about it: $1 to enter in 2000, $5 in 2001 (on the assumption that more data would be available), and $20 in 2002, with the result to be announced in 2003.[9]

As of July 25, 2000, 281 bets had been placed, ranging from 27,462 to 312,278; the mean and median were 67,006 and 61,302, respectively.[10] Bets kept coming, although they had to be signed in person, and the total pot eventually reached a life-changing $1200. Half was to go to the best 2000 estimate, a quarter each to the best 2001 and 2002 ones.

Regrettably, they didn't have an answer in time. Dr. Birney apologized for his "ridiculous hubris to think we'd have it nailed by 2003" but was finally persuaded to award the prize based on current best estimates.

The low bid won. Several numbers were proposed—26,000; 19,140; 24,500—but all were lower than the second-lowest entry; the estimate for award purposes was determined by esoteric means to be 21,000.[11]

Dr. Lee Rowen of Seattle came closest, with a bid of 25,947, made in 2001. Olivier Jaillon of France took the 2002 pot with 26,500. And the big winner was Dr. Paul Dear, who made a then-stunningly-low prediction

2.2 of 27,462 in 2000. How did he come to that figure? Three factors were involved:[12]

1. It was late at night and he had been drinking.
2. People's behavior at the time didn't seem so different from that of fruit flies, so double their 13,500 seemed ample.
3. As a Briton, he was used to writing dates in the form day/month/year, and he was born on the 27th of April, 1962.

Before we discuss genes, however, let's start with a reminder of what is essential about continuing the human species.

■ MAKING BABIES ■

CHILDREN ARE THE result of sperm and egg joining together and developing through a series of stages until eventually a baby is born.

Until very recently, this required the physical, sexual union of a male and female. Since the 1970s, however, the technique of in vitro fertilization (IVF) has made it possible for sperm and egg to be brought together in a laboratory, only later to be implanted in the woman who will carry the pregnancy to term.

Modern techniques have done wonders for infertile people. Women with blocked fallopian tubes, whose eggs cannot get into place to be fertilized, may now be able to have children (for a price—see **Chapter 5**). Men who produce little sperm, or sperm that does not move well, can often become fathers. In extreme cases, donors can supply the egg or the sperm; or a surrogate mother can even offer her womb and physically give birth to the baby.

All these technological developments, however, are just ways of helping sperm and egg to come together and develop. Or, in genetic terms, to enable the mingling and random assortment of the parents' genes. There may be some increased risk to the mothers, and possibly to the children, but parents who cannot otherwise have children willingly accept it, and society in general approves.

The technologies of assisted reproduction, discussed below, continue to develop in something close to a regulatory vacuum (see **Chapter 12**). They are often put into practice for people without prior animal studies, for example; and there remain questions about long-term safety, since the studies just have not been done. The possibility exists, therefore, that without public discussion—or even knowledge—some irresponsible, maverick practitioner may, for the first time in history, combine technologies, both standard and experimental, to produce babies that are *not* simply the result of joining sperm and egg.

That means, in genetic terms, changing a baby's genes—not as the random assortment that is the natural process of sexual reproduction, but deliberately, with a particular end in view.

■ THE BEAUTIFUL DREAM, FADING ■

THE THEORY BEHIND molecular biology, as formulated for the modern era in the 1950s, was elegantly simple. Genes were both the units of inheritance and the coding instructions according to which bodies—any bodies, of any species, including plants and bacteria—were built. The metaphors used varied; genes were spoken of as the blueprint for life, the operating manual, the encyclopedia, or (more recently) the system software.

Genes were strictly information workers. They instructed intermediaries to make the chemicals needed to do things—to grow, or perform repairs, or coordinate a mass of tissues to do something really complicated like sexual reproduction. The vital intermediate proteins received information from genes but could not pass information back the other way. Genes were passed on to descendants unchanged, give or take the occasional random mutation, which led Professor Richard Dworkin to write in his now notorious book *The Selfish Gene*, which was originally published in 1976:[13] "We are survival machines—robot vehicles blindly programmed to preserve the selfish molecules known as genes."

Even as a metaphor, this was *never* universally accepted among biologists. Prominent scientists, including Richard Lewontin, Ruth Hubbard, and Stephen J. Gould, for example, *always* objected to the "genocentrism" inherent in this approach. But there was enough truth there for the idea to be seductive—we all know people who look, and act, just like their parents—and the sheer simplicity of the concept that genes make us who we are was undeniably appealing.

The simplest working assumption was that there was a gene for every function—a gene for eye color, a gene for height, a gene for intelligence, and so on. This was always clearly an oversimplification (at least three pairs of genes are known to be involved in eye color, and there must be more[14]), but there's nothing necessarily wrong with that as a starting point for investigation. The way we think about gravity is an oversimplification, too, and that works.

Simple stories, however, can lead us down all manner of dubious paths; see, for example, **Chapter 4** (on stem cells) and **Chapter 8** (on eugenics). Once you accept the idea that a specific, identifiable *thing* causes us to be the way we are, it's a very short step to the idea of *doing* something to change the situation. That was the beautiful dream: We can know everything, and control everything. But reality is much more complicated.

Given that there are so many fewer human genes than once thought, it may be that the entire metaphor of genes as controlling devices needs adjustment. At least one writer has suggested that a gene should be looked at as a kind of Swiss Army knife, which can do any of several tasks, depending on how it is handled. That would imply that "fixing" things might be best accomplished by adjusting the controllers—except that we don't yet really know what and where they are.

What we do know more and more about are genes, partly because that is where we have been looking.

▪ GENES AND GENOMES ▪

GENES EXIST IN the chemical deoxyribonucleic acid (DNA), which is part of almost every cell of every organism. DNA is the same in each cell of any given organism, and virtually identical in all the cells of a given species. The genes in DNA are grouped into chromosomes (see **Box 2.3**), except for a tiny amount known as mitochondrial DNA (see **Box 2.4**), which is inherited separately.

2.3 THE NATURAL NUMBER OF CHROMOSOMES

OUR GENES, along with other elements, are contained in a number of chromosomes, which are specialized molecules. We have forty-six in all—twenty-three pairs, including the sex-specific pair. The other twenty-two are basically duplicated (one from each parent), as is the X chromosome for women; men have an X and a Y, while women have two Xs.

The duplication provides a partial fail-safe redundancy, which is often useful and incidentally explains why some diseases almost always affect men. But why forty-six? It's not a required number, though it certainly falls within a normal range.

Looking at other species, it becomes clear that there is no norm:

SPECIES	NUMBER OF CHROMOSOMES[15]
carp	104
chicken	78
horse	64
ape	48

2.3

Species	Number of Chromosomes[15]
human	46
mouse	40
guinea pig	16
rice	12
fruit fly	8
roundworm	6

That the number of chromosomes is variable has implications for genetic engineering. The first artificial human chromosome was reported as long ago as 1997.[16] What they promise—or threaten—is a more reliably consistent way of introducing artificial, or exotic, or at least replacement genes. If some of the would-be human engineers have their way, we may end up with forty-seven chromosomes yet.

2.4

THREE BIOLOGICAL PARENTS?

NOT QUITE all our DNA is in the chromosomes. A very tiny amount, 0.005 percent or so, with thirty-seven genes (maybe 0.15 percent of the total), exists separately in our cells.[17] This is in mitochondria, which are specialized, and essential, parts of a cell, and so it is called **mitochondrial DNA (mtDNA)**.

Mitochondrial DNA is inherited from the mother, as part of the outer egg. (To the great surprise of scientists, someone was discovered in 2002 to have inherited most of his from his father, but that seems to be very rare.[18]) It can of course go wrong—that is, contain a mutation that is damaging. About fifty diseases are thought to be the result of defective mitochondria.[19]

Some women may be infertile because of an mtDNA problem; at least some of the more adventurous fertility doctors suspect that. One proposed treatment was to use some mtDNA from someone else's healthy egg.[20] Doctors at a New Jersey clinic added parts of the healthy woman's egg into the patient's, along with her partner's sperm. The resulting

2.4 embryo—and child—had the usual mix of regular DNA from the two parents, but mtDNA provided by both women—three biological parents.

Unfortunately, two of the seventeen fetuses created this way had Turner syndrome and miscarried or were aborted.[21] That could have been a coincidence, but the Food and Drug Administration (FDA) jumped in fast and banned the practice.[22]

In late 2004, a British team applied for permission to experiment with foreign-mtDNA techniques.[23] They say they hope to avoid inherited diseases caused by mtDNA failures. Opponents are concerned that the process will put the child at great risk and be abused by older women (IVF success rates decline drastically with age and this might be a way of boosting them). They also see it as a significant step in the direction of Human GE, both technically and psychologically.

DNA itself consists of a huge number of only four units, which are known as bases and conventionally written as A, C, G, and T. These are roughly like the letters of the alphabet—string about 3 billion of them together in a particular order and you have a very, very long book, comprising a set of human DNA. In hard copy: about two hundred Manhattan phone books. Approximately three thousand of those units put together in the right sequence may form a gene, though some genes are very much larger than that; the dystrophin gene, mutations in which are associated with muscular dystrophy, is 2.4 million bases long.[24]

Some of this we know specifically as a result of the Human Genome Project (HGP). This was a massive undertaking that started in 1990 and formally ended in 2003 (see **Box 2.5**). It established an enormous amount of data, starting with the fact that the human genome consists of 3,164.7 million chemical nucleotide bases (units of DNA), of which less than 2 percent code for proteins, which in simple terms is how genes make things happen (see **Box 2.6** for some pointers to possible complications).

2.5 METAPHORS AND PERSONALITIES IN THE HGP

ON JUNE 26, 2000, with enormous fanfare, two rival groups of scientists jointly announced that they had "cracked the code" of the human genome.

The hoopla was somewhat premature (as everyone involved was well aware). What they announced was a "working draft," with about 85 percent of the contents in order and roughly half of that still needing double-checking.[25] One swooning journalist called it "an achievement so profound that no metaphor seemed able to capture it," but many tried.[26]

On April 14, 2003, with less fanfare but a certain amount of pomp, the leaders of the Human Genome Project (HGP) announced that their work was done, and the project was ending. It had lasted thirteen years, cost anywhere from half a billion to three billion dollars, depending on who's counting and what they include, and involved at least eighteen countries.[27]

Why then? It was the fiftieth anniversary of the announcement of the double-helix structure of DNA by Francis Crick and James Watson, who became the first Head of the HGP.[28] There remained, well, nagging little gaps, but this was considered a good enough time to declare victory and let the research continue without the structure of the HGP.

Already missing was the rival, commercial effort from Celera, which had been headed by Dr. Craig Venter. They chose to focus on drug development, basically because the HGP was giving away the data.

Venter took his money and became a genome philanthropist.[29] He later admitted, to the fascinated horror of most observers, that the genome whose sequence he and his colleagues published—the Celera one—was not that of some anonymous, homogenized group of donors, as had been assumed (and as the HGP one was).[30] It was that of J. Craig Venter himself.

The HGP didn't quite produce the promised Book of Life, but it's a workable metaphor. If each chromosome is a chapter, Nicholas Wade wrote in the *New York Times*:[31] "In the edition published [in 2003], small sections at the beginning, end and middle of each chapter are blank, along with some 400 assorted paragraphs whose text is missing, although the length of the missing passages is known."

That's getting close. As of this writing, researchers are into the last decimal places, and work continues on the individual chromosomes: Number 22 was completed first, in late 1999; the twelfth, Chromosome 5, was completed in September 2004, and marked the halfway point.[32]

At the same time, other genomes are being sequenced. Over 180 have been completely or largely determined.[33] What that does is enable comparisons across species, as well as between individuals, which can be a very useful research tool, because the similarities are often surprising (see **Box 2.6**).

2.6 **MYSTERIOUS JUNK**

PEOPLE HAVE A 3-gigabyte genome but only 2 percent of that consists of obviously working genes. The non-gene part has loosely been called **"junk DNA"** but that term seems less and less appropriate. All it really means is that we don't know what 98 percent of our genome does.

In 2004, a team led by David Haussler at the University of California, Santa Cruz, published a finding about "junk DNA" that, according to Haussler, "absolutely knocked me off my chair."[34] They ran a computer scan to compare the genomes of rats, mice, and people, and found 481 sections at least 200 units long that are absolutely identical—"ultraconserved regions." Those 481 are pretty much the same in other species, too:

SPECIES	% SIMILARITY OF ULTRA-CONSERVED REGIONS
human, rat, mouse	100
dog	99.2
chicken	95.7
fish	76.8

There are also many more, shorter, sections that are identical, or very nearly so; it may be that 5 percent of the genome is highly conserved.[35]

2.6   It's been 400 million years since humans and fish shared a common ancestor, long enough for significant changes to evolve, so what's going on here? One theory is that these sections must be essential, exactly as they are, or mutations would have crept in over the years.

Shockingly, however, another team of researchers managed to delete two huge chunks of the "junk DNA" of mice, including hundreds of conserved sequences—and it made no difference to the mice. They grew up apparently normal. Said Edward Rubin, who led the team:[36] "We were quite amazed."

No one knows what all this means, yet. At the very least, it seems that "junk" is the wrong word for most of our genome. And that we have a lot left to learn.

Each mouse has mouse DNA, and each of us has human DNA; but both are put together with the same "language"—that is, both consist of strings of ACGT—and both do have parts in common, quite significant parts. Almost all human genes have an equivalent in mice, though they are not identical. At the level of individual bases, the genomes are about 35 percent the same, but if you shuffle the mouse one around on a computer (which is cheating, but interesting), you can get it up to about 90 percent.[37]

As you might expect, apes have DNA more like people's than mice do (roughly 95 percent the same, by the most authoritative estimate) but human DNA is 99.9 percent identical for all of us.[38] The tiny differences, along with external factors like what we eat and do, are what make us each unique.

The critical *similarity* across species is the language. That's what makes it possible, at least in theory, to take genes from one organism and shove them into a quite different one with the hope that they will work. As long as they land in an appropriate place and don't get in the way of other, possibly important, functions. They don't always work (see, for example, **Box 2.7**), which is interesting, but if you can transfer them successfully, they will become part of the host genome.

2.7 THE MONKEY THAT DID NOT GLOW IN THE DARK

JELLYFISH GENES get spliced into all kinds of creatures, because the inserted gene is supposed to make tissues glow green under ultraviolet light and otherwise cause no harm. Aside from the crowd-pleasing element (see the glowing fish!) it makes a great test of principle. So that's what scientists at Oregon Health Sciences University did, to see if they could do genetic engineering on a monkey.[39]

After great efforts, one monkey was born (out of 126 embryos, created from 224 eggs) that carried that gene. But there was something strange: It did not glow. The gene was there but it was not working; in the correct parlance, it was not expressed.

Twins that miscarried did have fluorescent hair and fingernails—in the words of one breathless report, they "churned out glowing protein until they died"—but the live monkey did not.[40]

The experiment was good science; failure is a valid result. Still, it's a cautionary tale: Changes were made but the results were not as predicted. Was that a step forward toward Human GE, or a step backward? The answer seems to depend on your point of view.

■ SMUGGLING IN NEW GENES ■

IF YOU WANT TO affect something that is controlled by a gene, you have several alternatives, at least in theory. You could:

- ▶ replace the gene with a different one
- ▶ turn the gene off (or on) by introducing or affecting a regulator
- ▶ override the effect of the gene by adding or negating the protein it directs
- ▶ use an external appliance—such as wearing eyeglasses, putting on a sweater, supporting yourself with a walking stick, or driving a car

That last option is not really a joke; it's a reminder that we all extend (or override) our bodily capabilities all the time. Similarly, we can take medicine to counteract the internal effect of a gene's action. Or, we could try to change the gene itself.

One difficulty there is that our DNA is in essentially every cell in the body. We have trillions of copies of each gene. It's not quite as bad as that, since the ones we'd like to affect are the local ones that are acting in a particular part of the system, but it's still complex. And changing them all is not easy. Changing *one* is not necessarily easy!

Various methods are used to introduce "new" genes to cells. They are usually attached to some kind of virus, as a carrier, or **vector**. Sneaking into cells is just what viruses do; then they use the cell's own mechanism to replicate themselves. Several viruses have been modified to be harmless vectors.[41] Once inside the cell, the idea is that they will leave the new genetic material, and it will be integrated with the DNA already there. On the way, they have to evade the body's immune system, which is tricky if you are doing this directly on the patient (*in vivo*); an alternative can be to work on some of the patient's cells and then reinject them.

Exactly *where* within the cell's DNA the new gene ends up is also a problem—many locations could be fine, but some could disable another gene, or the mechanism that regulates another gene. Partly to avoid that, some researchers are working on developing artificial chromosomes, which could deliver a gene—in fact, many genes—without disrupting the DNA that is already there. There are problems with this, too: You still have to find a vector to introduce it, and that's harder because a whole chromosome is much bigger than the single snippet of DNA that makes up a gene.

That may make it all sound rather unlikely. But then, so is sexual reproduction, when you think about it: One sperm, out of millions, manages to fertilize an egg. Usually the process fails; but reproduction succeeds often enough. Introducing genes, however, really doesn't work as well as some people had hoped.

Unless you introduce them to the next generation. And that's where you really cross a major line.

■ CROSSING THE GERMLINE ■

THIS IS A distinction that is absolutely essential for understanding the controversies around Human GE: **somatic cells** are conceptually and practically different from **germ cells**:

▶ **"Germ cells"** are eggs, sperm, and the cells that make them; they pass on genes to offspring, and this shared lineage from parent to child is known as the **"germline"** of the organism or, more broadly, the species—the continuous inheritance from ancestor to descendant.

▶ **"Somatic"** comes from the Greek for "body" and may be used to contrast either with the psyche or with the germ cells, depending on context. In GE contexts, somatic interventions only affect one particular body.

There is a significant genetic way of expressing the difference: Somatic cells all have two sets of chromosomes, forty-six in all for humans (see **Box 2.3**); in technical parlance, they are *diploid*. Germ cells don't; they are *haploid*, which means they only have twenty-three chromosomes, one of each. (Sperm have either an X or a Y chromosome, plus 1–22; eggs always have an X and 1–22.) The reason for this is that they are designed to merge with a partner, in order to form the *zygote* that is the fertilized egg, and the beginning of reproductive development. Sperm and egg, the essential ingredients.

So, a tattoo is a somatic alteration; so is bodybuilding. If you work out, and increase your muscles, that makes no difference whatsoever to the genes you later pass on to your children. If you were able to use "gene doping" (see **Chapter 7**) to affect your muscles but not your germ cells, that would not affect your children either. If you doped your own germ cells or—more directly—the cells of a very early embryo during IVF treatment (see below), then *that* would indeed affect the germline. It would affect your children and their children, too.

In summary:

▶ **Somatic** genetic engineering affects some of the cells in a single body but is not passed on to future generations because it does not change the eggs or sperm.

▶ **Germline** genetic engineering does affect the eggs or sperm and is therefore passed on to future generations, who will carry the alterations in every cell of their bodies.

Every serious commentator—and regulator—considers the question of crossing the germline to be vitally important. A few people do want to do it (see **Chapter 10**), but governments everywhere, including the US, are worried enough by the prospect to forbid it. Some of the prohibitions are rather weak, bureaucratic rules that could easily be changed (see **Chapter 12**); many people argue that they should have the force not only of national law but also of international treaties.

Why are opinions about the germline so strong? Because breaching that barrier is essentially irrevocable. Humans, like any other species, have a **gene pool** that includes all the possible genes for our species. This changes, slowly and naturally, with mutations, most of which are probably harmful and do not survive—they are very likely to cause miscarriages, if development gets that far. A few do survive and become integrated into the gene pool. But deliberately changing that, especially when we do not know the full ramifications of the genetic mechanism, would be a very large step indeed. And if we alter or add a gene—an artificial one, or one from another species—it will stay there and become a permanent part of the gene pool. That's why some experts call the germline boundary:[42]

[T]he most consequential technological threshold in all of human history. . . . Crossing this threshold would irrevocably change the nature of human life and human society. It would destabilize human biology. It would put into play wholly unprecedented social, psychological and political forces that would feed back upon themselves with impacts quite beyond our ability to foresee, much less control.

Part of what makes this such a timely, as well as important, issue is that we are closer to being able to do it than we are to really knowing what we are doing. That in turn is partly because of the technological advances in fertility treatment—Assisted Reproduction Technologies (ART).

■ TECHNOLOGICAL REPRODUCTION ■

THE TECHNOLOGIES OF assisted reproduction were developed pragmatically. The goal was to help people, and the method was to see what worked. As the *industry* matured, some of the methods and motives of its practitioners became questionable (see **Chapter 5**), but the just-do-it attitude persisted, which has led to technical advances outpacing regulatory structures (see **Chapter 12**). The following is a very abbreviated description of the process.

In Vitro Fertilization (IVF) is the core technology. "In vitro" is the Latin for "in glass" and refers to the laboratory equipment. "Test tube baby" is actually a misnomer, as Louise Brown, the first of all, enjoyed pointing out to anyone who teased her: "It wasn't a test-tube, it was a petri dish."[43]

To be strictly accurate, in fertility contexts the name should include "Embryo Transfer" (thus, IVF-ET), but that's cumbersome. In practice, IVF covers retrieving an egg from a woman, and sperm from a man (that's usually easier), combining them in the lab, and, assuming all goes well, transferring the resulting embryo to a woman's uterus. (There are variations, listed in **Box**

5.3, that need not concern us, since they are either inappropriate for Human GE or limited in application.)

It is not absolutely necessary for a woman preparing for IVF to take hormone-stimulating drugs, but it is usual. They are controversial, and feminist activists insist that not enough is known about the risks—there have been reports of serious side effects, and there is a lack of adequate long-term safety data.[44] Under the influence of the drugs, the woman's system produces not one but, on average, 8–15 eggs at the same time.

Doctors retrieve the eggs surgically, separate them, one to a culture dish, and combine them with sperm. Ideally, that is fresh, though both previously frozen sperm and previously frozen eggs can at least in theory be used. (Frozen eggs are a very new phenomenon and in the largest known study 63 percent failed even to thaw successfully, and the success rate to birth was less than 2 percent.[45]) Each egg is placed in an incubator with about 10,000 sperm, under controlled conditions. With luck, fertilization will be detected in 12–20 hours.

Some men do not produce enough sperm, so an additional technique has been developed: **Intracytoplasmic Sperm Injection (ICSI)**. The practitioner selects one single sperm and injects it directly into the egg. This bypasses the usual selection methods—which sperm manages to survive the obstacle course of the layers surrounding the egg—but enables men with very low sperm counts to become fathers. It also guarantees that a particular single sperm is the only one involved, and no other residue remains; this can improve the accuracy of genetic tests (see below). ICSI is still regarded as experimental, and studies continue to see if its use is associated with later problems, but it has become a widely used technique.

A simpler alternative, of course, is to use someone else's sperm. Likewise, some women do not produce eggs successfully—but do have the physical capacity to sustain a pregnancy. They can be implanted with donor eggs, which has even been done with women who have gone through menopause. Conversely, if a woman can produce eggs but cannot sustain a pregnancy, then a surrogate mother can carry her developing baby to term. The techniques transfer relatively smoothly among these various possible combinations.

The embryos are left to grow for two or three days. At this very early stage, growth does not involve differentiation—all the cells remain identical to each other. About three days after fertilization, when each embryo has reached about the eight-cell stage, one or (usually) more are transferred to the uterus, while the rest are often frozen, for possible later use. (See **Chapter 5** for more discussion of the medical—and indeed financial—consequences for both parents and children.)

It is by no means certain that successful implantation and pregnancy will occur, as **Table 2.1** shows:

TABLE 2.1
ART SUCCESS RATES, 2001[46]
(USING FRESH, NON-DONOR EGGS)

PROCESS	NUMBER	PERCENT
Cycles started	80,864	100
Egg retrievals	69,515	86
Embryo transfer	65,363	81
Pregnancies	26,550	33
Live births	21,813	27

There are two ways of looking at this, contradictory but both true: The success rate is stunningly high—it works!—and worryingly low. Practitioners are therefore extremely interested in any technologies that will improve the rate. To some extent, they do this visually, picking what seem to be the most mobile sperm, and later the most healthy looking embryos. Some practitioners may also manipulate the incubation environment, in hopes of improving the implantation.

And now there is an important genetic test to add to their repertoire.

■ TESTING EMBRYOS ■

PREIMPLANTATION GENETIC DIAGNOSIS (PGD) relies on the fact that the early embryonic cells are identical. It seems to be possible, therefore, to remove one or two and test them, without damaging the remainder. There are still questions about the long-term effects of the procedure, but an estimated one thousand babies have been born after undergoing it.[47]

There isn't much time—a day or two—before the developing embryo has to be implanted, but the cell's DNA can be tested fairly quickly for certain specific mutations. If the results are favorable, then the IVF-ET procedure continues. If the result indicates a problem, that particular early embryo is discarded. Usually, several are in process, so if one "fails" another is likely to "pass" or at least to seem appropriate for implantation.

PGD can already detect, with a high degree of accuracy (90–99 percent):[48]

▶ errors in chromosomes that would lead to pregnancy failure, possibly failure even to implant, or to Down or Turner syndrome
▶ the presence or absence of a Y chromosome, indicating gender
▶ some single-gene disorders, such as Tay-Sachs disease, muscular dystrophy, and cystic fibrosis, and others for which genetic tests exist
▶ therefore, to some extent, whether a particular embryo is likely to become an appropriately matched tissue donor

Obviously, some of the possible uses are extremely controversial. PGD in itself does not *change* any genes, but it does allow an unprecedented ability to *select* them. For parents at risk of having a child with one of a few known specific heritable diseases, including Tay-Sachs and cystic fibrosis, this is a boon.

How it works is theoretically simple. Some people can be "carriers" of a genetic disease—such as Tay-Sachs—without suffering from it, since we have two versions of each gene; carriers have one good and one bad; sufferers have two bad. If both parents are carriers, then by very simple Mendelian genetics, their children have a 25 percent chance of getting two bad versions of the gene, a 50 percent chance of getting one bad and one good, and a 25 percent chance of getting two good versions. With PGD, the practitioner can select the healthiest embryo, and virtually ensure that the child won't have that disease.

That only really works for the rare single-gene diseases with definite on/off correlations with disease. Beyond them, the one thousand or so genetic tests that are available reveal only statistical data: With *this* gene or combination, the chances of getting *that* disease are *this* much higher. So there could be some difficult decisions to make—a 10 percent higher risk of this versus a 5 percent higher risk of that is a simple choice compared with the potential menu listing of trait possibilities, as PGD becomes even more accurate and computational improvements allow more tests to be performed simultaneously.

But if you are looking for a specific genetic combination—if you are either looking for a specific problem or you are actively performing genetic engineering and want to know if it worked—then it's an invaluable tool.

■ CLONING ■

IN THEORY, **cloning** follows some of the same steps as IVF, but with important differences. One is that it doesn't work very well, if at all (see **Chapter 3**), which is hardly surprising when you consider the process involved.

No sperm is involved. Instead, the egg is opened and its nucleus—the part that carries almost all of the DNA—is removed. In its place is put the nucleus from a somatic cell (that is, not a germ cell; see above) provided by the person or animal to be cloned. That is why the procedure is also called **Somatic Cell Nuclear Transfer (SCNT)**, a name that some scientists prefer to use; they say it's more accurate than "cloning," which can be used in different senses, but mostly they're trying to avoid the word, because people don't like cloning (see **Chapter 9**). SCNT just hasn't caught on; cloning it is.

There are various different techniques for accomplishing this strange fusion of donor nucleus and emptied egg.[49] First, the donor cell has to be made quiescent, so that the egg will accept it. (Some kinds of cells are naturally that way, making them better candidates.) Then its nucleus is either literally placed inside the egg, a particularly tricky piece of microsurgery, and then stimulated with chemicals, or placed right next to it and jolted into fusion by a little shot of electricity. The precise details vary, and research scientists have used trial and error in an attempt to create a "recipe," but it's still a mysterious and usually unsuccessful process.

A Korean team received global publicity in early 2004, because they had cloned human embryos, albeit without implanting them (see **Chapter 4**). They used fourteen different protocols, and the process was delicate enough that marginal changes in timing and so on made the difference; their eventual 26 percent success rate at starting the development of embryos (which are called blastocysts at that stage) was considered amazing.[50] Note from **Table 2.1** that this compares with about a 94 percent success rate in generating transferable embryos from eggs and sperm in standard IVF practice.

The result of all this is an embryo that can, in simple theory, be implanted and grow through pregnancy to become a cloned baby. There is some evidence, discussed in **Chapter 3**, that this process can never be done reliably. In part this may be due to the lack of sperm, or perhaps to inadvertent damage created during the fusion process.

If, however, such a child existed, he or she would be almost genetically identical to the donor from whose somatic cell the clone was derived. Not quite, because of the mitochondrial DNA (see **Box 2.4**), which lies outside the nucleus and is supplied by the egg. If a woman were to clone herself, with her own egg and her own somatic cell, then both the mitochondrial and the nuclear DNA would indeed be identical, but the uterine environment inevitably would not be—that's a variable that is undeniably important, but not easily quantified.

The most plausible candidates to perform cloning are professional fertility specialists. Dr. Antinori and Dr. Zavos, both profiled in **Chapter 3**, claim that they have developed expertise in the delicate art of selecting and nurturing embryos, and there may be some truth to that. Most scientists, however, remain deeply skeptical about their abilities.

■ USING CLONING TECHNOLOGY FOR HUMAN GE ■

IF HUMAN CLONING could be made reliable, or even reasonably predictable, it would almost certainly become part of the foundational technology for human germline genetic engineering—the making of what the popular press calls "designer babies." So would **stem cells**, which are the subject of **Chapter 4** and described there in more detail. For the moment, it is sufficient to say that embryonic stem cells (ESCs) exist in the first eight weeks of pregnancy, and are capable of developing into almost any kind of cell in the body.

Conceptually, here's the process:[51]

1. Use IVF to create a single-cell embryo (zygote).
2. Allow it to develop until ESCs have formed (about five days), then isolate the stem cells.

3. Culture the stem cells (in order to have plenty to work with) and introduce the new genes.
4. Grow the modified stem cells and test that they do have the new genes.
5. Take a new egg, remove the nucleus (as in cloning), and replace it with the nucleus of a genetically modified stem cell.
6. Let that genetically modified embryo develop, and implant it in a woman's uterus.

The big advantage of this procedure is that you only have to successfully modify *one cell*—and it will, theoretically, develop into a GE baby. You have as many chances as you have stem cells to work with. Moreover, you can check that the new genes have been smuggled into that cell before beginning the cloning process, and with PGD you can double-check before implanting the embryo. In something with as many variables as GE, quality control is essential.

Of course, it may not work. It's all a tricky and rather unpredictable business, which is why Dr. John Burn, professor of clinical genetics at Newcastle University, England, has warned about the potential dangers of trying to use human genome information to "improve" the species: "It would be rather like kicking the television to get a better picture at the moment."[52]

But some people really do want to try. In fact, they seem almost to look on the possibilities of genetic tampering with something that suspiciously resembles religious fervor (see **Box 2.8**).

2.8   **SCIENTIFIC THEOLOGY, OR THEOLOGICAL SCIENCE**

THE FOUNDERS OF modern molecular biology always thought big. When James Watson (see **Chapter 10**) and Francis Crick figured out the shape of DNA, the famous double helix for which they shared the Nobel

2.8 Prize with Maurice Wilkins, they went down to the local pub to cele-
brate. As they entered, Crick announced loudly:[53] "We've found the
secret of life."

A few years later, in 1958, Crick laid out the canonical form of a vital
belief, that proteins cannot pass information to nucleic acid, RNA or DNA,
or to other proteins. He called this, with a typical touch of whimsy, the
Central Dogma of Molecular Biology. It is often simplified as "DNA to
RNA to protein," to which Crick strenuously objected, since he had not
eliminated the possibility of communication from RNA to DNA.[54]

Several people have commented on the similarities between the
"faith" of biotech and the faith of religion. Leigh Turner, for example,
wrote in *Nature Biotechnology* in 2004:[55]

> Biotech is not just an assemblage of research programs and tech-
> niques. In a scientific and technological era, biotech also offers a
> surrogate religious framework for many individuals. We might
> want to explore the dangers associated with turning biotech into
> a belief system. . . . The religion of biotech needs to be challenged
> by debunkers and skeptics.

Of course, that's how science is supposed to work—new facts
crowding out old theories. That's not actually what happens, much of
the time. Researchers who produce unexpected results should expect
scrutiny; they shouldn't face attacks on their personal integrity. It happens, though, as the well-documented cases of Arpad Pusztai in Britain
and Ignacio Chapela in California attest.[56] Both produced findings
inconvenient to the GE food industry, and both were assaulted with a
relish that seems to go beyond the merely self-interested and into the
realms of Inquisition.

The Central Dogma itself is now under assault. A few "heretics" such
as Barry Commoner have been hammering away at it for a while.[57]
Another is the British-based Dr. Mae-Wan Ho, who writes in a piece
titled, with not a little glee, "Death of the Central Dogma":[58] "It is not so
much that we need a new theory to replace the central dogma; it is
more important than that. We need a new way of knowing and being

2.8 organisms that will prevent us from mistaking organisms for instruments and machines. That's the real challenge."

Let's hope she doesn't get burned at the stake.

■ THE NEXT FRONTIER? ■

OVERSIMPLIFICATION IS the enemy of accuracy. This chapter has *not* told you how genetics works, and it has *not* told you that genetics does not work. It has told you far less than you need to know to perform science, and considerably more than you need to make social and political judgments about how to use science.

There is more detail on certain scientific topics spread through the rest of the book, especially in Chapters 3–7. Along with them are examples of "stories" that are not exactly correct but are used to "sell" lines of research; funding is usually a problem for scientists in the modern world of business-financed universities.[59] There are also plenty of examples of sensationalistic or otherwise misleading media coverage—reader beware!

What there is not is a discussion of **proteomics**, which is often called the next frontier of research for molecular biology. The word itself is a bastardized concoction of "protein" and "genomics" first coined in about 1995.[60] It refers to the study of the proteins in an organism, and their relationship to genes and to physiology. In a sense, it's switching the center of attention away from genes (of which there are so disappointingly few) toward proteins, of which there are maybe a million.

As an area for research, it's huge. Whereas the genome of a person is set and constant, the proteome is in continual flux, and by some estimates there may be as many as 20 *billion* chemical interactions to figure out.[61] That'll keep scientists busy for a while.

And maybe "protein therapy" will be more to the point than "gene therapy." Or maybe not. There is so much left to learn.

■ FURTHER READING ■

Free Documents from the Web

"The Spurious Foundation of Genetic Engineering," by Barry Commoner, *Harper's*, 02/02, is available at www.mindfully.org/GE/GE4/DNA-Myth-CommonerFeb02.htm, with updates and comments on readers' responses at criticalgenetics.org.

"Death of the Central Dogma," by Dr. Mae-Wan Ho, 09/03/04, is at the Institute of Science in Society (ISIS) website, www.i-sis.org.uk/DCD.php, with links to other articles; memberships and books are for sale.

The US Department of Energy, which coordinated the Human Genome Project along with the National Institutes of Health and partners such as the UK Wellcome Trust and others, still maintains an informative website at www.ornl.gov/sci/techresources/Human_Genome/home.shtml.

The leading scientific journals *Nature*, nature.com, and *Science*, sciencemag.org, both keep some important content free to the wider public, though you may have to register. *Nature*'s "Web Focuses" are available from nature.com/biotech. *Science*'s "Essays on Science and Society" are at www.sciencemag.org/feature/data/150essay.shl.

Books

Ruth Hubbard and Elijah Wald, *Exploding the Gene Myth*, Beacon Press, 2nd ed. 1999; Professor Emerita of Biology at Harvard, Hubbard knows both science and people, while Wald makes sure the rest of us can understand.

Richard C. Lewontin, *Biology As Ideology: The Doctrine of DNA*, Harper-Perennial, 1993; "the most subversive book to be published in English in 1993," by the Alexander Agassiz Professor of Zoology and Professor of Biology at Harvard University.

Richard C. Lewontin, *The Triple Helix: Gene, Organism, and Environment*, Harvard University Press, 2000.

Jeffrey M. Smith, *Seeds of Deception*, Yes! Books, 2003; focuses on GE food, but much of the science is relevant.

James D. Watson, *The Double Helix*, originally published by Atheneum, 1968; Norton Critical Edition (including reviews and essays), New York, 1980.

■ ENDNOTES ■

1 "GFP Bunny" is a "transgenic artwork" created by, or for, Eduardo Kas in 2000; see www.ekac.org/gfpbunny.html.

2 Michael Hopkin, "Salmon Give Birth to Trout," *Nature*, 08/04/04

3 Dr. David Whitehouse, "Mars 2 – Earth 0," BBC 12/06/99

4 "Human Gene Number Slashed," BBC, 10/20/04

5 Nicholas Wade, "Human Genome Now Appears More Complicated After All," *New York Times*, 08/24/01

6 Kevin Davies, *Cracking the Genome*, The Free Press, New York, 2001

7 ibid., p. 103

8 Nicole Johnston, "Rice Genome Rising," *The Scientist*, 03/01/04

9 Nicholas Wade, "Gene Sweepstakes Ends, but Winner May Well Be Wrong," *New York Times*, 06/03/03

10 Statistics downloaded from www.ensembl.org, where Dr. Birney worked, on 07/25/00

11 Wade, 2003, op. cit.

12 Wade, 2003, op. cit.

13 Richard Dawkins, *The Selfish Gene*, 2nd ed., Oxford University Press, 1989

14 See www.athro.com/evo/gen/inherit1.html or www.seps.org/cvoracle/faq/eyecolor.html, for example, for more detailed discussion.

15 See www.ornl.gov/sci/techresources/Human_Genome/faq/compgen.shtml and www.en.wikipedia.org/wiki/Chromosomes

16 J. Travis, "Human Artificial Chromosome Created," *Science News Online*, 04/05/97

17 16,569 base-pairs, compared with 3.1 billion; www.cellbio.utmb.edu/cellbio/mitoch2.htm

18 "Mitochondria Can Be Inherited from Both Parents," *New Scientist*, 08/23/02

19 Antony Barnett and Robin McKie, "Babies With Three Parents Ahead," *Observer*, 10/17/04

20 Jason A. Barritt, Carol A. Brenner, Henry E. Malter, and Jacques Cohen, "Mitochondria in Human Offspring Derived from Ooplasmic Transplantation: Brief Communication," *Human Reproduction*, 03/01

21 Rick Weiss, "Pioneering Fertility Technique Resulted in Abnormal Fetuses Team Failed to Disclose Genetic Disorder in Science Report," *Washington Post*, 05/18/01

22 Rick Weiss, "FDA to Regulate Certain Fertilization Procedures," *Washington Post*, 07/11/01

23 Barnett and McKie, op. cit.

24 See the Human Genome Project website, www.ornl.gov/sci/tech resources/Human_Genome/project/info.shtml.

25 Rick Weiss and Justin Gillis, "Teams Finish Mapping Human DNA," *Washington Post*, 06/27/00

26 Richard Saltus, "Rival Teams Present Rough Findings in Efforts to Decode Human Genome," *Boston Globe*, 06/27/00

27 The Department of Energy, the bureaucratic base of the HGP, maintains an informative website at www.ornl.gov/sci/techresources/Human _Genome/home.shtml.

28 See www.nature.com/nature/dna50/index.html for the double celebration.

29 The Genome News Network (www.genomenewsnetwork.org) is a project of the J. Craig Venter Institute.

30 Nicholas Wade, "Scientist Reveals Genome Secret: It's His," *New York Times*, 08/24/01; www.nytimes.com/2002/04/27/science/27GENO.html

31 ibid.

32 The Genome News Network lists these reports at www.genomenews network.org/categories/index/genome/chromosomes.php.

33 The 180-plus (as of March 2005) genomes to have been sequenced are listed, with links, at www.genomenewsnetwork.org/resources/ sequenced_genomes/genome_guide_p1.shtml.

34 Helen Pearson, " 'Junk' DNA Reveals Vital Role," *Nature*, 05/07/04; www.nature.com/nsu/040503/040503-9.html. The original paper was by Bejerano, G. et al. and published in *Science*.

35 Dr. Mae-Wan Ho, "Are Ultra-conserved Elements Indispensable?" Press release from the Institute of Science in Society (ISIS), 09/16/04; www.i-sis.org.uk/AUEI.php

36 Sylvia Pagán Westphal, "Life Goes On Without 'Vital' DNA," *New Scientist*, 06/03/04

37 Ben Shouse, "UCSC Researchers Play Key Role in Mapping of Mouse Genome," *Santa Cruz Sentinel*, 12/05/02

38 Roy J. Britten, "Divergence Between Samples of Chimpanzee and Human DNA Sequences Is 5%, Counting Indels," *Proceedings of the National Academy of Sciences*, 10/04/02. This finding supersedes the formerly accepted figure of 98.5% similarity between chimps and people, but it is still an approximation that masks both similarities and differences.

39 "OHSU Researchers Produce First Genetically Modified Monkey," Oregon Health Sciences University Press Release, 01/11/01. The formal report was in *Science*, 01/12/01, pp. 309–312. A good summary is: Rick Weiss, "Scientists Create First Genetically Altered Monkey," *Washington Post*, 01/12/04

40 Steve Sternberg, "Monkey Glows Green, for Human Benefit," *USA Today*, 01/12/01

41 See www.ornl.gov/sci/techresources/Human_Genome/medicine/gene therapy.shtml

42 "The Threshold Challenge of the New Human Genetic Technologies," Center for Genetics and Society, overview at www.genetics-and -society.org/overview/threshold.html

43 "From Miracle Baby to Regular Teen," *People*, 02/07/94

44 Our Bodies Ourselves (OBOS), also known as the Boston Women's Health Book Collective (BWHBC), has been particularly active on the fertility drug issue; see www.ourbodiesourselves.org/lupron.htm.

45 Kristen Philipkoski, "Frozen Eggs Showing Promise," *Wired*, 09/13/04

46 US Department of Health and Human Services, *2001 Assisted Reproductive Technology Success Rates*, p. 15; available at www.cdc.gov/reproductivehealth/art.htm

47 *Preimplantation Genetic Diagnosis: A Discussion of Challenges, Concerns, and Preliminary Policy Options Related to the Genetic Testing of Human Embryos*, a report of the Genetics and Public Policy Center at Johns Hopkins University, 01/08/04; available as a 34-page pdf or series of web pages at www.dnapolicy.org/policy/pgdOptions.jhtml.

48 The President's Council on Bioethics, *Reproduction and Responsibility: The Regulation of New Biotechnologies*, Washington, D.C., March 2004, pp. 92–94; available from www.bioethics.gov.

49 The Stanford University website has a more thorough description than most, of three methods, at www.stanford.edu/~eclipse9/sts129/cloning/methods.html

50 Gina Kolata, "Cloning Creates Human Embryos," *New York Times*, 02/12/04

51 There is a useful illustration of this process, among others, at www.genetics-and-society.org/technologies/igm/science.html.

52 BBC World Service, 02/11/01

53 James D. Watson, *The Double Helix*, originally published by Atheneum, 1968; Norton Critical Edition, New York, 1980, p. 115

54 Francis Crick, "Central Dogma of Molecular Biology," *Nature*, 08/08/70

55 Leigh Turner, "Biotechnology as Religion," *Nature Biotechnology*, 22, 659–660 (2004)

56 Jeffrey M. Smith, *Seeds of Deception*, Yes! Books, Iowa, 2003, especially pp. 5–44 and 221–227.

57 Barry Commoner, "The spurious foundation of genetic engineering," *Harper's* 02/02

58 Mae-Wan Ho, "Death of the Central Dogma," Institute of Science in Society (ISIS), 09/03/04

59 Eyal Press and Jennifer Washburn, "The Kept University," *Atlantic Monthly*, 03/00

60 By Marc R. Wilkins, according to Marie McCullough, "Study of Proteins Leads to Medical Breakthroughs," *Philadelphia Inquirer*, 04/23/04

61 Justin Gillis, "Big Buildup to the Genome," *Washington Post*, 02/09/01

3

CLOWNS AND COWBOYS IN THE CLONING CIRCUS

■ INTRODUCTION ■

C LONING HAS BECOME a freak show, an easy headline, and a source of cheap laughs at the "clowns" or "cowboys" claiming to do it to humans. But the circus atmosphere conceals some very serious issues: Like a magician's patter, the carnival surrounding the self-promoting self-proclaimed cloners serves to distract us from much of what is really going on.

People may be getting hurt, and worse may be to come.

Moreover, the arguments against cloning include many of the most important arguments against all human genetic engineering. The reasons that drive a few people to encourage cloning—which are not necessarily the arguments they put forward to justify it—are also worth examining. Cloning may not be the worst possible example of human GE, but it is the most immediate. And it is still not illegal in most of the United States.

■ WHAT IS CLONING? ■

A CLONE IS, essentially, an artificial copy. The word is the Greek for "twig" and was adopted about a hundred years ago to describe plants grown from cuttings, grafts, and the like, rather than from their usual—sexual—method of reproduction.[1] In modern terms, the key fact is that the DNA of the offspring is effectively identical to that of the source (see **Chapter 2** for the minor qualifiers to that statement, and for the technology in general).

From there, the usage developed, you might almost say organically, to cover various obviously related actions (see **Box 3.1**). It then spread to encompass almost any kind of copying— the graphics program Photoshop includes a "clone stamp" tool for duplicating parts of an image; derivative rock bands can be derided as "heavy metal clones" by critics; politicians can be called "Reagan clones," and so on.

3.1

DEFINITIONS

THIS BOOK USES "cloning" in all three of its common biological senses, to refer to:

- cell duplication
- embryo creation for research purposes, which may include the extraction of embryonic stem cells, and
- the making of lambs, pups, calves, kids, piglets, kittens, bunnies, babies, and other offspring

A **"clone"** is the product of cloning, usually in the third sense.

These overlapping usages very rarely cause confusion, but as insurance qualifiers may be added, thus:

Some of those who want to do **research cloning** have noticed that the public overwhelming rejects **reproductive cloning** (see **Chapter 9**). They would like, therefore, to use a completely different term. They prefer to call the process **SCNT**, or **Somatic Cell Nuclear Transfer**, which refers to the method by which cloning is usually done (see **Chapter 2**).

In practice, however, research cloning advocates are mostly stuck with that name, or else a term no one likes: **therapeutic cloning**. Those in favor of it don't like to say "cloning" and those who oppose it say it's not exactly therapeutic; at best, it's a process that may lead to something that might eventually be used in a medical treatment (see **Chapter 4**).

The President's Council on Bioethics, bending over backward to be fair, devoted twenty pages and fifty-five hundred words of Human Cloning

and Human Dignity: An Ethical Enquiry to an exhaustive discussion of ter-
minology, and finally settled on the clumsy but accurate phrases **"cloning-
to-produce-children"** and **"cloning-for-biomedical-research."**

People really in the know, however, when they are talking about
cloning to make babies, refer casually to **"repro"** as in, "Why can't we at
least ban repro?"

These later, looser, metaphorical usages led to some confu-
sion, shown by this anonymous e-mail question, cited in an
Associated Press report:[2] "Let's say I am 33 and I want a clone.
Is my clone also 33 or are they born as a baby?"

No, clones do not spring to life as full-grown adults. Just as
a plant can grow from a cutting, so a clone would have to
develop. Your clone (if one could be made) would not be your
age, would not have your memories, and would not have any
skills you had learned. Your clone would be the same gender as
you, since that is determined absolutely by your DNA, and at
thirty-three might well look more like you did at thirty-three
than you will at sixty-seven or so, but even that's not certain;
environment has a lot to do with development, beginning with
the environment of the mother's womb.

An adult human clone, if one were to exist, would be the
result of a childhood following from a pregnancy, begun by the
implantation of an embryo, using methods developed in fertility treatments (see **Chapter 5**). It is the *creation* of that embryo
that would be the most obvious difference. That would be the
result of an egg being hollowed out, having some tissue inserted
into it, being stimulated by electrical and/or chemical shocks
and, essentially, being fooled into developing like a fertilized egg
(see **Chapter 2**).

That sounds implausible, doesn't it? It is. The idea of cloning
animals has intrigued scientists since at least the 1920s.
Repeated experiments, however, failed, or at best achieved
some partial success like producing tadpoles that did not
develop into frogs. By 1984, the distinguished researchers

James McGrath and Davor Solter could write in *Science* that,[3] "The cloning of mammals, by simple nuclear transfer, is biologically impossible."

They were wrong, of course (see **Box 3.2**), but not by much. Cloning is still extremely difficult at best, still has not been achieved in several species with which scientists have tried hard, and possibly has never produced a fully healthy result (see **Box 3.3**).

3.2

THE MOST FAMOUS CLONE

THE FIRST MAMMAL to be cloned from an adult cell was a **sheep**. Known as "Dolly," she was born on July 5, 1996, at the Roslin Institute in Scotland, and announced on February 27, 1997.[4] She was killed by injection on February 14, 2003, suffering from lung cancer and arthritis, at the unusually early age of six years old. There was no provable connection between her illness and her status as a clone, though premature death "is typical of cloned animals" and that breed of sheep "can live to 11 or 12 years of age."[5]

The attraction of cloning to the researchers who achieved it was that it would let them perform consistently repeatable genetic engineering. They wanted to genetically modify farm animals (cattle were the real target), but the process is unpredictable and often unsuccessful. If they could succeed once, and then clone the result, this could, they thought, have significant implications.

From the start, cloning technology was intended to enable genetic engineering.

3.3

MAMMAL CLONING SUCCESS RATES

SOME SCIENTISTS SUGGEST that all clones may be defective,[6] but by late 2004 eleven species of mammals had been reported as successfully cloned. In the following table, the first seven entries include all births for each species reported before July 2002, when the most authoritative chronicler stopped keeping track, and none thereafter; the rest are from contemporary reports.[7]

Species	Eggs Used	Clones	Efficiency	First Success
sheep	>2,350	13	0.6%	1996
mice	>19,000	141	0.7%	1997
cattle	>12,820	181	1.4%	1997
goats	>2,000	30	1.5%	1998
pigs	>8,500	16	0.2%	2000
cat	87	1[8]	1.1%	2002
rabbits	~2,000	4	0.2%	2002
mules	334	3	0.9%	2003
horse	841	1	0.1%	2003
rats	>130	2	<1.5%	2003
deer	?	1	?	2003

By July, 2002, fifty papers had been published in peer-reviewed scientific journals, describing sixty-eight experiments, involving about 45,000 eggs, 552 "live births" that did not live long and 386 surviving clones. More individuals have been cloned since, but with similarly low success rates.

Failed experiments are not always published, so the true percentage could be even lower. Births as a percentage of embryos transferred are of course higher, but in only two of the sixty-eight studies did they exceed 25 percent efficiency, both with cattle; at least twenty-nine were under 5 percent.

Other Breeds and Species

A kind of wild cow, a **banteng**, was also cloned in 2003 (using a surrogate cow), as was a European **mouflon**, a kind of sheep. A wild ox, a **gaur**, was cloned, the sole survivor of 692 eggs from which thirty-two embryos were implanted, but it died less than two days after its birth in 2001.[9]

Monkeys have been "twinned" by splitting and then duplicating the very early embryo (a form of cellular cloning), which led to a birth in 2000, but despite over one thousand attempts there have been no verified reports of monkey cloning in the accepted sense of using an adult's DNA.[10] In late 2004, it was reported that cloned monkey embryos had been created, but none of those led to a pregnancy that lasted longer than a month.[11] This did, however, serve to cast doubt on a suggestion made by the same respected researcher the year before that it might be impossible to clone primates.[12] (He used a technique

3·3

pioneered with human eggs, discussed in **Box 4.9**, though the human embryos were not implanted.)

Cat cloning has become a business (see **Chapter 10**), though there are proposals to ban it. By March 2005, six cats had been cloned and two sold, at $50,000 each. The same company hopes to achieve its original goal of cloning a **dog** in 2005, although they call them "the most difficult species to clone."[13] (Why that should be is not clear, but certainly results do vary among species.) There have also been thus far unsuccessful attempts to clone **pandas**, **chickens**, the extinct **Tasmanian tiger**, the even longer extinct **woolly mammoth** (using an elephant surrogate), and probably other species.[14]

Mice and one bull have been **"recloned"**—that is, clones have themselves been cloned. It took 358 eggs to get the second-generation bull clone; attempts at a third generation were abandoned after 248 attempts.[15] Mice, however, have been cloned unto at least six generations.[16]

Finally, the first successful cloning of insects—**fruit flies**—was reported in November 2004.[17] It took eight hundred attempts to produce five cloned flies. The plan was to improve mammalian success rates through experimenting with flies, though some experts were extremely doubtful that this would work.

Cloning is not a way to mass-produce soldiers (or athletes, or geniuses), since each one would have to develop individually, in a particular woman, subject to her unique uterine conditions at the time, and then in a particular family and social environment, at a particular school, and so on. (That's part of what distinguishes clones from twins, see **Box 3.4**.)

THE DIFFERENCE BETWEEN CLONES AND TWINS

3-4

IDENTICAL TWINS arise as a result of a fertilized egg (or very early embryo) splitting, to form two embryos, both of which develop. They always have the same basic DNA as each other, and also the same womb environment and mitochondrial DNA (mtDNA; see **Chapter 2**), which comes from the mother during pregnancy.

3-4

Fraternal twins develop from different eggs; they share mtDNA but not overall DNA.

A **clone** would have the same basic DNA as an older person, but different mtDNA, because of the different mother; in that, clones differ from both identical and fraternal twins. Moreover, clones are manufactured, whereas twins arise during the natural process of reproduction; the methods, relationships, and purposes are radically different.

If a single-cell pre-embryo—a fertilized egg—were artificially duplicated, and both embryos developed, the process would be cloning, in a different sense (see **Box 3.1**), and the result would be artificial twins. Similarly, a very early embryo can deliberately be split, as happens naturally when identical twins form.

Calling twins "natural clones" or clones "delayed twins" is either simplistic or propaganda designed to make the process seem familiar and thus acceptable.

■ HAS ANYONE CLONED A HUMAN? ■

No. Several people have claimed to produce a human clone or to know of one (see **Boxes 3.5–3.9**), but none have proved it. Given modern DNA testing, proof would be easy; given how hard it is to clone animals, the burden of proof is definitely on those who claim that a human clone exists.

It is worth noting that most of the main *advocates* of human cloning have called the noisiest claims, by the Raelians in 2002–4, a hoax:

> This is just a great big stunt, and you fellows are being played for every bit of media coverage they can get.
> —*Randolfe Wicker, Human Cloning Foundation*[18]

Anyone can stand up and look pretty in front of a camera and say: "We have got it, we've cloned a baby," but where's the beef? In science, you don't cut corners, you don't tell

untruths, you don't come up with these things without proof.
. . . This is too ugly a show . . . it's an ugly circus.
—*Dr. Panayiotis Zavos*[19]

Despite this, and despite overwhelming disapproval of the idea, a significant number of Americans do think that a human clone has already been born in secret—30 percent in one 2002 poll thought this "very likely" and another 26 percent thought it "somewhat likely," even though 89 percent in the same poll considered cloning "unacceptable."[20]

Why do so many Americans think human cloning has been done? Possibly because of confusion with the whole issue of making stem cells from cloned *embryos* (see **Chapter 4**). Cloned embryos have indeed been made, at least twice, but they either failed to develop or were used as sources of stem cells.[21] There are other reports, not formally published in peer-reviewed journals, of cloned embryos that were not intended for pregnancies and may have been deliberately destroyed to prevent any possibility of their being implanted.[22] But births? No.

3·5

HUMAN CLONING CLAIMS 1: RORVIK

THE FIRST UNSUBSTANTIATED claim of human cloning to achieve wide publicity was a 1978 book, In His Image, by the experienced science journalist David **Rorvik**, which purported to describe a clone born two years earlier.

Rorvik and his publisher, J.B. Lippincott, were sued by a scientist mentioned in the book, who proved that Rorvik had asked him how to clone five months *after* the clone was supposedly born, and a year after Rorvik had supposedly agreed to help the alleged parent, "Max." The court ruled that the book was "a hoax and a fraud," and the publisher settled the suit for a reported $100,000. The publisher had made $730,000 and the author $390,000.[23] Rorvik still has not admitted that he made it up.

3.6 HUMAN CLONING CLAIMS 2: THE RAELIANS

THE RAELIANS are a cult led by the former Claude Vorilhon, who now calls himself Rael and claims that "life on earth was created scientifically in laboratories by extraterrestrials" (who told him about it).[24] They caused a front page frenzy by announcing that a clone had been born on December 26, 2002, a girl called—what a surprise—Eve. The sole parent was said to be a thirty-one-year-old American whose husband was infertile.

Clonaid is a "brand name" (it's not technically a company) set up by Rael to sell cloning.[25] CEO Brigitte Boisselier originally said they would produce the baby for DNA testing, and the results would be available in a week. Six days later, however, Rael told CNN that he had advised Boisselier that "if there is any risk that this baby is taken away from the family, it is better to lose your credibility; don't do the testing."[26]

Sure enough, when someone dragged Clonaid into court less than a month later, they said they were no longer in contact with the family, who were believed to be living in Israel. No testing has been reported.

To diminishing media returns, Clonaid announced clones born to Dutch, Japanese, Saudi and Brazilian parents. There was a pause, then they claimed on their website that eight more clones had been born in February and March 2004, in "Australia, Mexico (two of them), Brazil, Spain, Italy, England, and Hong Kong."[27]

No reference to the 2004 claims could be found in a database search of mainstream US media conducted that June. In September 2004, however, more was revealed, in a sense: Three female Raelians, including Boisellier's adult daughter, posed nude for *Playboy's* October issue. The Movement proudly issued a press release.

■ WHAT IS WRONG WITH CLONING? EVERYTHING! ■

IN EVERY POLL, huge majorities of people agree that human cloning is unacceptable (see **Chapter 9**). Specific reasons are discussed in the next section, but one that resonates with many people is extremely simple: **Cloning is repugnant.**

This statement either strikes a chord with you, or doesn't. No one will be convinced who doesn't already agree; but those who think it makes sense often like to know they are not alone. And they are not: The reasons most people give for opposing cloning tend to be equally vague.

Consider the poll summarized in **Table 3.1**. Researchers asked an open-ended question—so these responses were volunteered, not picked from a menu—and then put the answers into categories, grouped roughly by topic. (The table retains the original categories but regroups them somewhat differently than the original report.) Evidently, three-quarters of the general public and half of all working biologists, perhaps more, essentially think cloning is wrong because it's wrong.

TABLE 3.1
MORAL AND ETHICAL CONCERNS
ABOUT HUMAN CLONING[28]

"WHAT ARE YOUR MAIN MORAL AND ETHICAL CONCERNS
REGARDING HUMAN CLONING?"

	GENERAL PUBLIC	WORKING BIOLOGISTS
A. REPUGNANCE		
Playing God/Interfering with Gods' work	30	9
Religion (general)	4	2
Against the Bible	1	0
Against cloning (general)	17	7
Messing with nature/environment	10	7
Ethically/morally wrong	7	10
Potential abuse of technology/science	4	14
subtotal	73	49

	GENERAL PUBLIC	WORKING BIOLOGISTS
B. CONSEQUENCES		
Clone would not have soul/ be individual	6	3
Emotional/psychological problems for person cloned	4	2
Against selective breeding	5	7
subtotal	15	12
C. OTHER ISSUES		
Dangerous/risky for all	3	4
Lack of control	2	4
Not enough testing yet	2	9
Uncertainty about how research will be used	1	8
Uncertainty about who's deciding about research	1	4
subtotal	9	29
Total	97	90

An expressed concern about "potential abuse," for example, begs the question, "Why is it abuse?" The answer "I'm against cloning in general" is even less responsive. Experimental safety is a valid ethical concern, discussed below, but some of the "other issues" seem to have little to do with the question, and may indicate an amorphous feeling of unease, especially among those scientists who don't like appealing to religion. In fact, it is hard to avoid the conclusion that biologists don't have a much better idea of why they oppose cloning than anyone else, they just express their views somewhat differently (see also **Box 4.3** for a survey of geneticists, who overwhelmingly oppose reproductive cloning).[29]

Just because almost everyone agrees on something doesn't mean they're right; they could simply share a prejudice. Still, it does seem that the normal attitude to cloning reflects something deep in our nature, comparable to our attitude to incest. There are many logical reasons to ban incest, but concerns about familial power dynamics or the genetic condition of inbred offspring seem somehow inadequate to provoke the usual revulsion. Moreover, intellectual arguments alone do not explain why both promiscuous monkeys and Japanese quail avoid incest, nor why close family members of humans tend, according to researchers, to smell both recognizable and bad to us.[30] (That has been proposed as an incest-avoidance mechanism.) Incest, it seems reasonable to say, is just wrong.

So is cloning. We need not be shy to say so. As Leon Kass (see **Box 11.6**) wrote in a groundbreaking essay,[31] "repugnance is the emotional expression of deep wisdom, beyond reason's power fully to articulate it."

■ SPECIFIC REASONS NOT TO TRY CLONING PEOPLE ■

DIFFERENT PEOPLE stress different arguments against cloning, and use somewhat different language to make their points, but a survey of eight lists of secular objections quickly reveals the outlines of a consensus.[32] The reasons given can almost all be grouped under four general topics:

- ▶ safety, on several different grounds
- ▶ eugenics and what cloning might lead to
- ▶ objectification and commodification
- ▶ personal, familial, and social relationship issues

Each of the eight sources mentioned both safety and eugenics (or social justice, or something similar), sometimes linked with potential technologies; almost all mentioned commodification and/or manufacture; and at least five referred, generally more than once, to questions of individuality, uniqueness, familial, and social relationships, etc. (Some also mentioned respect for nature, religious arguments, concerns about "murdering embryos," or other issues that might fall generally under the category of "repugnance" considered above.)

CLONING IS NOT SAFE

We *know* cloning is not safe because of the experiments done to animals. (The morality of animal cloning is not considered here, though a significant majority of the public oppose the practice; see **Chapter 9**.) Animal success rates remain consistently terrible (see **Box 3.3**)—and that is on the dubious assumption that the cloned beasts are healthy; most, if not all, are in fact sick, if not obviously deformed.

Cloning expert Dr. Jose Cibelli, a strong supporter of human cloning research for medical reasons but *not* for reproduction, has explained the odds for cow cloning, step by step:[33]

- 30 of every 100 eggs will die in the petri dish
- 15 of the 70 remaining will develop enough to be transferred, to 7 cows
- 4 of the 7 cows may become pregnant
- 2 of the 4 fetuses will die early in the pregnancy
- 1 of the 2 remaining will die just before birth
- 1 clone might be born

No wonder Cibelli stresses, in the strongest terms, that: "Whoever wants to attempt [human cloning] is a criminal and should be penalized."

Moreover, some scientists suggest that cloning can *never* be predictably and reliably safe, especially for primates. Early attempts to clone monkeys resulted in "genetic chaos. . . , with chromosomes distributed almost at random," though later studies seem less conclusive.[34] Some mouse studies suggest that cloned embryos are hampered in development, compared with either naturally conceived ones or those created through IVF and similar techniques.[35] The researchers involved do not claim to know why, but they too warn against human cloning.

There are more general indications that messenger RNA from sperm may be involved in the timing of early development—in other words, that sperm is needed for more than simply fertilizing the egg.[36] Some development clearly does happen without it, at least in some species, but this finding may help explain the low success rate. That would suggest that **cloning may be permanently unpredictable** in *all* mammals.

Despite the evidence, and the virtually unanimous scorn of mainstream scientists, some cloning advocates claim that their experience in human fertility clinics means they are more likely to get it right. How could they find out? Only by trying. That runs straight up against the long-established ethical principle that risky **human experiments** can only be performed with the consent of the subject. (See **Chapter 12** for discussion of the Nuremberg Code, Helsinki Declaration, and Belmont Report.) The clone cannot possibly give consent; can the mother? Many would argue that she cannot both be in her right mind and consent to put herself and any possible clone at such a level of risk. Cloners counter that IVF treatment, which is now routine, would never have been developed under this standard; they may be right, but opposition then was only theoretical, and vanished with success (see **Chapter 9**). To repeat, we *know* that success rates in *every* species where cloning has been attempted are *terrible*, and the worst experiences of all are with the species most like us. Bucking that trend is the definition of unacceptable risk.

A longer-term safety issue might be **genetic diversity**. Cloning by definition repeats a genome, so it avoids new genetic combinations. That lessens variety, and thus may lessen the chances of some member of a species having built-in protection against certain diseases. This has been a significant issue in agriculture for years. For example, at least 85 percent of the 7,098 apple varieties known in nineteenth-century America were extinct by 1990, and an even higher proportion of the 2,683 varieties of pears; modern seed banks, which are intended to maintain diversity, came much too late for most of these.[37] How does cloning come into this? Well, bulls are being cloned for their sperm. . . . People are not, and mass production is absurd, but there are realistic fears of other long-term consequences, which we turn to next.

CLONING AND EUGENICS

Cloning is by its nature eugenic, since it involves the deliberate production of someone with the "right" genome. (See **Chapter 8** for a full discussion of eugenics.) That in itself is a reason to question, if not ban, the practice, though cloning is not going to make wholesale changes to society or the gene pool; it's strictly retail and bound to be small-scale. What is worrying is the contribution that cloning could make to encouraging and developing widespread eugenics.

Technologically, techniques associated with cloning are practically essential to the modern techno-eugenic effort (see **Chapter 2**). This does *not* mean the making of babies, exactly, but the duplication of very early embryos, probably using stem cells (see **Chapter 4**), in order to maximize the chances of modification. Work done to make cloning more successful almost automatically makes the chances of **human genetic engineering** greater. Indeed, this is cited as one of the reasons *for* cloning by some of the few people who want to do it. The mindset that considers cloning appropriate is precisely the mindset that embraces the idea of Human GE.

The technology is important, but the **psychology** around cloning is perhaps even more significant. To say that someone is foolish because they want to clone themselves, or a relative, or a famous person is to miss the point: It's foolish to want to clone *anyone*. The essence of eugenics is the idea that any of us might be capable of identifying the "best" humans; that's exactly what any cloner does, in their own definition of what is desirable. Not only will they fail (because of the unpredictability of the final result) but they are foolish to try. *Every* such attempt falls back into the mists of prejudice (see **Chapter 8**). It's bound to. Each of us is the result of random recombinations, and it's not just repugnance that insists that's the way it should be, it's also that we simply do not know better, and there is no evidence beyond our own hubris that we ever will.

3.7

HUMAN CLONING CLAIMS 3: ANTINORI

DR. SEVERINO ANTINORI is an eminent Italian specialist in human fertility. He runs an IVF clinic in Rome and has published many scientific papers. He pioneered controversial postmenopausal pregnancies, some with women in their sixties, and developed innovative techniques for maturing human sperm in a test tube with material from rats' testicles.[38]

On May 5, 2004, he said that at least three human clones had been born, and he had been involved, but only as an advisor.[39] This did not cause much of a stir, perhaps because editors felt burned by the Raelians (see **Box 3.6**), or else because he had made so many claims before:

- In February 2001, he predicted that he would have cloned a human "within the year."[40]
- In April 2002, he said, "One woman . . . is eight weeks pregnant."[41]
- Later that month, he told Italian TV of two cloned pregnancies in Russia and one in an "Islamic state." His former partner, however, said Antinori had "no clones, no laboratory, no patients and no doctors to help him."[42]
- In June 2002, the word was, "The pregnant now is five. Until now, no birth. No delivery. Miscarriage, only one."[43]
- In December, at least a month after the original claimed baby would have been born, he announced that the first clone would be born in Belgrade in January 2003.[44]
- The following July, gliding gracefully past a question about his previous statements, he claimed that "documented proof with

3·7

a photograph of the five-month-old cloned foetus will be published shortly in a medical journal."[45] A search of the journal ten months later found no recent articles by Antinori, and no relevant ones that mentioned him.

Antinori is said to have brought at least thirty-six libel suits, and claims to have won six against the Vatican.[46]

CLONING TREATS PEOPLE AS OBJECTS

Cloning is a form of manufacture, and as such implies that the cloned creature is an object, an artifact. The making of clones also touches on the commercial market, where raw materials and production technology are bought or rented. And the entire procedure is one more—extreme—example of the medicalization of natural processes.

All these aspects—**commodification, objectification, medicalization,** and **manufacture**—have been the subjects of more detailed analysis than we have space for here. Commodification, for example, is a relatively new word for the process of turning natural relationships (such as sexual love) into financial relationships; reducing "the family relation into a mere money relation."[47] Some capitalist theorists think this is a good thing (see **Box 5.2**); but many conservatives, including such distinguished bioethicists as Leon Kass and Daniel Callahan, agree with Karl Marx and most modern progressives that it is not. No one can really deny, however, that cloning is an example of commodification.

The materials used for cloning are likely, in the US, to be bought and sold. (This is illegal in Europe, Canada, Australia, and Japan, at least.[48]) Women are already being offered significant sums (occasionally as much as $50,000; see **Box 5.2**) to sell their eggs, and any human cloning is likely to require hundreds of eggs (see **Chapter 4**). When Advanced Cell

Technologies (ACT) experimented with cloning (though not to make babies), their ethics advisory board wrestled long and hard with this issue and eventually settled on a **payment** of $4,000. The scientists subsequently complained that they couldn't get enough eggs, which suggests that the price may rise and poor women may be tempted even further to take dangerous drugs and undergo surgery for money. Again, some people are not troubled by this; most social justice activists, however, consider it appalling.

Moreover, there is a scientific objection to the point of view that encourages objectification. Any medical procedure to some extent does involve taking part of a person and looking at it objectively, in isolation—as an arm, say, or a heart. But that is both a useful and a dangerous tendency, as a generation of feminists in particular have reminded us.[49] Medicine, and science in general, require that we consider the whole and its surroundings and interrelationships, as well as the simple sum of its obvious parts. As the study cited above about messenger RNA from sperm reminds us, reproduction is extremely complex—and a *process* rather than an object.

RELATIONSHIP ISSUES

The last set of objections to cloning concern interpersonal relationships. What is the relationship between donor and clone? Who exactly are the parents? What would the expectations of the donor be for the clone, and how would that affect the clone's upbringing, psyche, and subsequent existence? Vice versa, what would be the effect on the donor? How would society as a whole be affected?

The counter to all these questions is that each of them represents no more than an exaggeration of concerns that are perfectly familiar. Parents always have expectations of their children; adopted children, and their parents, have particular issues that are somewhat similar to those cited; society has adapted to the modern reality of divorce. This is at least partly true, but modern reproductive technologies may be making some problems worse. Most official views are cautiously optimistic, like this World Health Organization survey's conclusion:[50] "Although existing knowledge about the impact of assisted reproduction for parenting and child development does not give undue cause for concern, there remain many unanswered questions about the consequences of creating families in this way."

But some of the subjective reports are much more disturbing. There have been reports of children born from sperm donation having psychological difficulties. One English woman, for example, says she always felt strange, although she was forty when she found out that:[51]

> [M]y father was a glass jar with a blob of sperm in it. My father doesn't have a face, or a name and he wasn't even a one-night stand. If my mum had had an affair at least there would have been sex and lust, something human rather than something so cold, scientific and clinical. My parents never even met. How weird is that? I still feel like a freak, a fake. I don't feel I know who I am anymore. . . . I was only made to assuage my parents' reproductive vanity.

How much worse would those feelings be for a clone?

Perhaps these relationship issues represent no more than another version of repugnance; but it seems significant that so many of those who have considered this subject in detail, from the California Cloning Commission to the President's Council on Bioethics, not to mention liberal and progressive campaigners, consider them to be serious questions. The pro-cloners, however, don't.

3.8 HUMAN CLONING CLAIMS 4: ZAVOS

DR. PANAYIOTIS (PANOS) ZAVOS is more modest in his claims than Dr. Antinori (his former partner), the Raelians (whom he condemns), or Rorvik. Some experts regard him as more plausible, "a scientist who may even be able to deliver on his promises," but he has not produced a cloned baby.[52] Instead, he courts publicity for the incremental steps, but without giving proof.

In April 2003, he published a picture said to be of a four-day-old cloned embryo, but the promised peer-reviewed analysis has not appeared.[53]

In January 2004, he announced at a press conference in London that he had implanted a cloned embryo into a thirty-five-year-old woman, though he said there was a 30 percent chance that the implantation would be successful. Two weeks later, he announced that the client was not pregnant.

Dr. Zavos, a PhD and former professor of animal sciences—not a physician—does run a fertility clinic in Kentucky, with his wife, who is certified in obstetrics and gynecology. The clinic, however, has consistently been listed as "nonreporting" by the US Centers for Disease Control, so its success rate is unknown.[54] He has claimed to do "hundreds of normal embryo transfers every month."[55]

Among his other businesses is a "Home Semen Analysis Kit" to help men discover if they are infertile.[56] The front page of his website touts his interviews on national TV and testimony to Congress.[57] Returning a California writer's call in early 2001, he immediately began pitching the screenplay potential of his story.[58] He remains very cooperative with the press, although wide open to comments like this one, from a profile in the London Daily Telegraph:[59] "In the past 10 years he has been sacked by a hospital for 'unethical and illegal behavior,' fined for improperly labelling discarded sperm samples at his private lab and ordered to pay a former employee about $500,000 for malicious prosecution."

Zavos seems magnificently unfazed. He is reported in the British press to charge £35,000 (over $60,000) for an initial consultation about cloning.[60]

■ JUSTIFICATIONS GIVEN FOR CLONING PEOPLE ■

WHATEVER MOST OF US feel about cloning, a few people (not many; see **Chapter 9** for the opinion polls) do think it's acceptable. They may have unacknowledged motives, discussed below, but they do of course present justifications.

Some of their arguments focus on claims that cloning technology will lead to the elimination of disease, a fountain of eternal youth, or even safer breast implants. These are mostly calls for human genetic engineering, which many people consider to be a major reason *not* to allow cloning. There are also arguments against banning cloning on essentially libertarian grounds, but they generally assume rather than argue that cloning is desirable.

Arguments made specifically in favor of reproductive human cloning include:[61]

- *Cloning will let me have a particular child*, for example to replace someone who died. This is factually inaccurate, as well as being a lousy reason to have a baby. Cloning will not duplicate a dead child (or anyone else), since personality, character, abilities, and temperament, as well as physique, are all subject to environmental constraints.
- *Cloning will let me avoid having a child with an inherited disease.* This is unnecessary at best. Some object to the practice on principle, but early testing (before implanting a regular IVF embryo) can ensure that an embryo is free of a specific genetic disease, without cloning.
- *Cloning will let me grow spare parts for transplants.* This is grotesque, if it means growing an entire "designer donor" (could they turn you down?) and quite fantastical if it means growing a kidney in a petri dish.

▶ ***Cloning will allow a completely infertile person to reproduce.*** But modern technology, though expensive (as cloning would be) can enable almost everyone to become a parent. (See also next.)

▶ ***Cloning will enable a woman to have a daughter with no man involved.*** If cloning were to work—which is extremely doubtful—this would technically be true. The traditional alternatives (adoption and sperm donation) remain, however, and this previously unimaginable possibility seems inadequate to counterbalance all the arguments against cloning.

Some would-be cloners don't agree with all of those arguments anyway. The Italian fertility expert and cloning enthusiast, Dr. Severino Antinori (see **Box 3.7**), insisted in 2001 that he had refused to clone a woman's dead son ("No way"), and would only try in cases where the man could produce no sperm.[62] He gave as an example an American couple where the man had lost both testicles in a car accident: "Ask any woman. If her man could not produce even the first stage of sperm, would she prefer to be implanted with the sperm of a stranger, and bring up a stranger's child, or would she rather have something which bears the genetic imprint of the man she loves?"

Note that, in this account, a child with the wife's normal complement of DNA is "a stranger's" if she uses donated sperm, and a child with *none* of the wife's DNA is far preferable to one with none of the husband's. So, a woman is presumably no more than a vessel for reproducing her husband. Many would disagree.

■ OTHER POSSIBLE MOTIVATIONS ■

SOME PEOPLE really, really want to do cloning. Billionaire John Sperling financed a program at Texas A&M to clone his beloved dead dog. When that didn't work, the program morphed into a cat-cloning business (see **Chapter 10**).

It is hard to avoid the suspicion that this urge to clone is the repugnance argument turned inside out and upside down. Curiosity, money, fame—they seem secondary to this odd desire to just do it. Endangered species! *Extinct* species! Never mind that the ecosystem of which they were part has been destroyed, we'll correct our techno-errors with techno-fixes. . . . This is irrational exuberance, and does not seem to respond to logic.

Some scientists are working on animal cloning with the aim of producing drugs or even transplantable organs for people; those are at least worthy goals, even if society decides to avoid those means. Others may have an understandable interest in cloning on a technical level; but doing it is another matter. And on humans? Forget it.

3.9   DR. SEED, AND THOSE WHO WOULD CLONE THE KING

THE WONDERFULLY NAMED Dr. Richard Seed announced in early 1998 that he was going to produce five hundred clones a year, starting with himself and his wife. He showed little evidence of the resources or expertise required, though he had been involved in some legitimate work on embryo transfer techniques; his doctorate is in physics.[63] He was described as "quite serious and seriously nutty" by Lee Silver (see **Chapter 10**), himself no enemy of cloning.[64]

The publicity Seed gathered was one reason that President Clinton called for a law banning cloning. This in turn led to the following statement:[65] "Americans for Cloning Elvis (ACE) neither supports nor disapproves of Dr. Seed's plans. Americans for Cloning Elvis (ACE), however, is concerned that Dr. Seed's plans will derail its goal of cloning Elvis."

They know of a piece of toenail and a wart, they say. They had at least 2,637 signatures on June 17, 2004. Why bother? Elvis lives.

■ SO WHY HASN'T CLONING
BEEN BANNED YET? ■

IN MUCH OF the world human reproductive cloning is illegal already. About fifty countries, including Canada and most of Europe, have specifically banned it, as well as at least eleven US states (see **Chapter 12**).

That said, there is still no federal law forbidding it in the US (though the FDA has a de facto moratorium, at least, in place). This is certainly a failure of the political system, given the enormous majorities in favor of such legislation (**Chapter 9**). Partly, it is attributable to the entanglement of this issue with abortion politics (see **Chapter 4**), which in turn is to some extent deliberately and very cynically encouraged by some biotechnologists.

That might seem odd, but there are serious scientists who do still want to develop cloning techniques. They accept that cloning is not now safe, but hope to improve it. A few of them actually call for a moratorium, rather than a ban, on reproductive cloning, while others openly admit they would prefer to have no law against it than a law they don't like.[66] And that's what they've got—no law.

Here's how it happens. Some—not all—religious conservatives oppose all research that involves embryos. Rather than working to isolate them and limit their influence, lobbyists for the biotechnology industry seem to restrict their own activities to defeating legislation they don't like; they do not encourage regulation they could live with. Said one frustrated congressional staffer about the biotech lobbyists: "We asked for their advice on more than one occasion. But they were not willing to offer any constructive proposal. They gave the impression that they were going to oppose anything."[67]

Reproductive cloning has become a sort of test case for Human GE. It's not one that most GE advocates wanted, since they cannot really argue that there are medical reasons for it. Nor is it the one that opponents would necessarily have chosen, since it does not exactly involve altering the human germline, though the technologies involved are indeed relevant (see **Chapter 2**). To some extent, both camps view the issue as something of a distraction, but it was forced on us by a few mavericks.

The arguments in favor of it are very flimsy; the arguments against are very powerful; and it is extremely unpopular. The failure to ban cloning on the federal level would be a terrible example of the shortcomings of our political system. Banning cloning, however, would set a useful precedent for the control of potentially dangerous technologies.

■ FURTHER READING ■

Free Documents from the Web

The Human Genome Project "Cloning Fact Sheet" is a substantial overview of the history, science and ethics of cloning, with well-selected links arranged by topic, at www.ornl.gov/sci/techresources/Human_Genome/ elsi/cloning.shtml.

The President's Council on Bioethics, *Human Cloning and Human Dignity: An Ethical Inquiry*, is a 340-page report, available as a single pdf (3.3mb) or a series of web pages from www.bioethics.gov/topics/cloning_ index.html.

The British Broadcasting Corporation (BBC) hosts a series of public-service web pages under the rubric "Religion & Ethics" that includes lists of arguments on both sides of the issue and some excellent links; www.bbc.co.uk/ religion/ethics/cloning/clones.shtml.

Pro-cloners, would-be cloners and self-proclaimed cloners can be found at www.humancloning.org, www.clonerights.com, www.zavos.org, www.rael.org and www.clonaid.com.

Books

Leon R. Kass and James Q. Wilson, *The Ethics of Human Cloning*, The AEI Press, 1998; an uncommonly good-looking little book, which includes Kass's classic essay, "The Wisdom of Repugnance."

M. L. Rantala and Arthur J. Milgram, eds, *Cloning: For and Against*, Open Court, 1999; a collection of 54 essays with a wide range of points of view.

Gina Kolata, *Clone: The Road to Dolly and the Path Ahead*, William Morrow and Company, 1998

David Rorvik, *In His Image: The Cloning of Man*, J. B. Lippincott Co., 1978, is generally assumed to be science-fiction.

▪ ENDNOTES ▪

1 *Oxford English Dictionary*, 2nd ed, 1989. The earliest written record is from 1903, in *Science*, which spelled it 'clon'; but C. L. Pollard, in the same journal two years later, suggested adding the final 'e,' presumably to ensure that people pronounced it properly.

2 Lauran Neergaard, "Americans Don't Understand Cloning," AP, 04/02/01

3 *Science*, 12/14/84

4 *Nature* 385: 810–813. She was named by technicians after the buxom Ms. Parton, since the adult cell used for the cloning derived from a sheep's breast. It was a stroke of marketing genius, all the better for being accidental. BBC, 05/30/00.

5 *Nature*, 02/18/03, in an obituary archly subtitled "Celebrity clone dies of drug overdose."

6 Defective clones: BBC, 08/07/03; *New York Times*, 02/17/04

7 The entries for sheep, mice, cattle, goats, pigs, and rabbits are based on data compiled by Lesley Paterson, Roslin Institute, last revised July 2002; www.ri.bbsrc.ac.uk/public/cloning.html. Roslin calculated "cloning efficiency" by comparing "live births" rather than "number survived" with eggs used, which gives slightly higher percentages.

8 Genetic Savings and Clone sold its second cloned kitten in February 2005 and promptly lowered the price to a mere $32,000. See www.savingsandclone.com for the company's position and www.NoPetCloning.org for details of proposed legislation to ban the practice in California.

9 Mouflon: BBC, 09/04/03. Gaur: BBC, 10/08/00; *Washington Post*, 01/13/01

10 *New Scientist* 12/12/01 and 04/08/02, *Chicago Tribune*, 06/25/02, *Times of India*, 10/21/02

11 Helen Pearson, "Biologists Come Close to Cloning Primates," *Nature*, 10/21/04

12 *New Scientist* 04/10/03, citing *Science*, vol. 300, p. 297

13 Reuters, 10/09/04

14 www.ornl.gov/sci/techresources/Human_Genome/elsi/cloning.shtml; BBC, 11/27/02; *New York Times*, 05/28/02; *Economist*, 08/07/03

15 AP, 05/23/04, citing *Nature Biotechnology*
16 *Nature*, 407:318–9, 2000
17 "Scientists Report First Insect Clones," *Atlanta Journal-Constitution*, 11/04/04
18 www.clonerights.com/history_of_the_movement.htm
19 *Financial Times*, 01/06/03
20 Fox News, 02/15/02
21 *Washington Post*, 11/26/01; *Science*, 02/13/04
22 BBC, 12/16/98; *New Scientist*, 04/09/03
23 Gina Kolata, *Clone: The Road to Dolly, and the Path Ahead*, New York: William Morrow and Company, 1998, p. 118. See www.msu.edu/~millettf/writings/cloning.html for a defense of Rorvik, which cites an article by him in *Omni*, 04/12/01.
24 See the Raelian website, www.rael.org.
25 CBS News, 06/02/03
26 CNN, 01/03/03
27 Their website, www.clonaid.com, noted that "The cloned child born in Australia was not conceived within Australia and therefore adheres to Australian cloning laws."
28 Survey conducted by KRC Research for the Biotechnology Industry Organization, 06/04
29 *Genetic Engineering News*, Vol. 24, No. 4, 02/15/04.
30 Frans de Waal and Frans Lanting, *Bonobo: The Forgotten Ape*, University of California Press, 1997; P. Bateson, "Preferences for Cousins in Japanese Quail," *Nature* 295: 236, 1982; *New Scientist*, 08/22/02
31 *The New Republic*, 06/02/97
32 The sources are: the President's Council on Bioethics; the BBC; the Center for Genetics and Society; the Council for Responsible Genetics, and also the article they reposted by Sophia Kolehmainen; Our Bodies Ourselves; Friends of the Earth; and the Foundation on Economic Trends.
33 www.drkoop.com/newsdetail/93/8009645.html
34 See notes 11 and 12.
35 "Glimpse Into Cloned Embryos Reveals Problems," *New Scientist*, 06/30/04
36 *Nature*, 05/13/04; Reuters, 05/12/04
37 *New Internationalist*, 10/90
38 *People*, 01/24/94; London *Daily Telegraph*, 07/19/95; BBC, 03/07/99
39 Reuters, 05/05/04
40 The London *Times*, 02/20/01
41 *The Australian* 04/08/02
42 London *Daily Telegraph*, 04/27/02
43 *Chicago Tribune*, 6/25/02
44 London *Daily Telegraph*, 12/18/02
45 *Times of India*, 07/29/03, 07/30/03

46 *Scientific American*, 04/02/02; *Chicago Tribune*, 6/25/02
47 The word "commodification" dates to 1975, according to the Oxford English Dictionary; the concept, and quotation, are from the 1848 *Communist Manifesto* by Karl Marx and Friedrich Engels.
48 Hawley Fogg-Davis, "She Works Hard for the Money? Reproductive Labor and Racial Desire," available at http://ptw.uchicago.edu/FoggDavis02.pdf
49 Notably, the Boston Women's Health Book Collective, creators of *Our Bodies, Ourselves*; see www.ourbodiesourselves.org.
50 Effy Vayena, Patrick J. Rowe and P. David Griffin, eds., *Current Practices and Controversies in Assisted Reproduction*, report of a meeting on "Medical, Ethical and Social Aspects of Assisted Reproduction" held at WHO Headquarters in Geneva, Switzerland, September 17–21, 2001
51 London *Observer*, 01/20/02
52 *Scotsman*, 10/14/03
53 *New Scientist*, 04/09/03
54 www.cdc.gov/reproductivehealth/art.htm
55 *Scotsman*, 10/14/03
56 www.sementests.com, linked from zavos.org
57 www.zavos.org
58 Friends, I was that writer. Zavos was extremely pleasant, and took the news that I live nowhere near Hollywood very well.
59 London *Daily Telegraph*, 02/14/04; I silently changed "about £300,000" back to dollars, as reported in the same article, and Americanized "behaviour."
60 *Scotsman*, 10/14/03
61 It is surprisingly difficult to find a list of coherent arguments in favor of cloning to make human babies put together by people who believe them. (Cloning opponents often put together lists of arguments to attack, as here.) See www.humancloning.org/benefits.php, especially the bottom of the page, for as good a try as you are likely to find, though it omits the "man-free" argument.
62 London *Sunday Times*, 01/28/01
63 *Christian Science Monitor*, 01/13/98; *New Scientist*, 01/17/98
64 Lee M. Silver, *Remaking Eden*, Bard, 1998, p. 306
65 www.geocities.com/americansforcloningelvis
66 Dr. Paul Berg, for example, testified in Sacramento (03/20/02) that he would prefer a 10-year moratorium on reproductive cloning. Dr. David Gollaher explicitly preferred no law at all to a "bad law" at a Stanford, CA panel discussion, 03/10/02; Dr. Berg and others present voiced agreement.
67 Asif Ismail, "Lobbying, Old-Time Politics Block Legislation on Human Cloning," Center for Public Integrity, 02/15/01

4

THE SCIENCE AND POLITICS OF STEM CELLS

■

■

■ INTRODUCTION ■

S TEM CELLS SEEM, at first glance, to represent the most benign side of human genetic technologies. Advocates of embryonic stem cell research in particular promise miracle cures, and the loudest opposition seems to come from fundamentalist conservatives, against whom any Democrat, Green, or progressive will line up instinctively.

Unfortunately, it's not that simple.

Embryonic stem cells definitely represent a fascinating area for research, but that does not necessarily mean that medical treatments using them will be available soon. Dr. Jose Cibelli, a staunch supporter, says he has "no idea" when therapies may be available. Dr. Jeffrey Rothstein, who is also optimistic, says, "It's foolish to give timelines." Dr. Peter Andrews, a British researcher with thirty years of experience, insists:[1] "I caution, there is an awful lot of hype. Using stem cells—in whatever form—routinely is quite a long way away. I say, I seriously doubt there will be any treatments [using stem cells] for anything for fifty years." Even Dr. Irving Weissman, an active campaigner for funding embryonic stem cell research, when pushed, talks about twenty years—and says he'd be shocked if we have "a salable product" in five years.[2] This avenue of research is a long-term proposition.

Concern about the unlimited and unregulated use of embryonic stem cells, especially, is by no means limited to conservatives. Many feminists are specifically troubled by the prospect of women being exploited for their eggs—and especially by the cavalier way this concern is often overlooked. Judy Norsigian of *Our Bodies, Ourselves* testified before Congress that:[3]

Of particular concern to us is the lack of adequate long-term safety data on the super-ovulating drugs that women have to take in order to provide the eggs for embryo cloning. . . . [Furthermore,] it is highly likely that many women will become repeat donors, and that there would be massive expansion in the use of women as paid "egg producers." We know nothing about the health risks of such repeat donations.

The simplistic claims of many advocates for research on embryonic stem cells do indeed require millions of human eggs (see **Box 4.4**). The burden for supplying these, and taking the risks involved in the process, would surely land on poor women, selling their tissues for money.

There is also a different aspect of fairness: Who benefits? Any developed therapies of the kind described below would be expensive. An analysis by Rockefeller University scientist Peter Mombaerts in *Proceedings of the National Academy of Sciences* noted the large number of eggs required, and the complexity of the procedure, and estimated that "to generate a set of customized [embryonic stem] cell lines for an individual, the budget for the human oocyte material [eggs] alone would be $100,000–200,000."[4]

Then there is the "slippery slope" toward technological abuse: As described in **Chapter 2**, some of the technology that would need to be developed to make "therapeutic cloning" a reality would also be crucial to the development and application of germline—inheritable—Human GE.

Stem cells do *not* have to have anything much to do with the worrying side of human genetic engineering—but in practice that's where some of their advocates are heading. Their use does *not* have to entail cloning—but the political push does. And overall the stem cell debate does *not* have to be intertwined with abortion politics, though for some people some parts of it always will be.

Politically, in the US, cloning and stem cells have been joined in a way that has not only made it impossible to come to enough agreement to pass a national law of any kind on the subject—either subject—but may also have buried some questions we *should* be addressing. Therefore, both the **science** of stem cells—what is known and what we hope to learn—and the **politics** of the issue, which are something of a case study in how *not* to make good decisions, are subjects of this chapter.

Is political stalemate inevitable? No. Americans need look no further than Canada to find a model that could be adapted for use south of the border, one that respects conflicting rights and enables valuable research to proceed (see **Chapter 12**).

That is why many progressives, environmentalists, and feminists were pushing in 2002 for a formal moratorium on embryo cloning, specifically to ensure that appropriate rules were put in place first. President Bush's Council on Bioethics also recommended a moratorium in 2002, by a majority, some of whom favored a permanent ban. At one point, this seemed like a promising way forward, but it stalled in the face of opposition both from conservatives who thought it inadequate and from some scientists who thought it too restrictive.

In the US, we have at present the worst of all possible worlds—one in which the unscrupulous are free to do anything they can pay for; while some medical researchers cannot get government funding for their work. Embryos aren't protected. Basic science is hindered. No one is happy.

■ WHAT ARE STEM CELLS? ■

THERE ARE MANY kinds of cells in our bodies—liver cells, tooth cells, blood cells, and so on. Every one of them carries a complete set of our particular DNA, but only uses a tiny fraction of it, because every one has a particular function. They are all specialists, and when they reproduce (which not all do) they can only make more of the same—liver cells make liver cells, and so on.

STEM CELLS BY FUNCTION

ONE USEFUL WAY of classifying stem cells is according to their potential for producing other kinds of cells (another is by their origin; see **Box 4.2**). Three categories are generally used:

> **totipotent**—able to produce any kind of human cell at all, including an entire embryo; only true of the very earliest stage after fertilization; from the Latin for "able to become the whole thing"
>
> **pluripotent**—able to produce almost any kind of cell (except totipotent cells and some cells of the tissue layer surrounding the embryo); form inside the embryonic cell mass after a few days' development; "able to become more things"
>
> **multipotent**—able to produce a limited range of specialized cells, for example the various kinds of blood cell; from later development and presumably the adult tissues; the Latin root ("many things") may be a bit misleading—"several things" might be better

The term "produce" should be understood as "lead to the production of"—remember, all this science is simplified. Quite how limited multipotent stem cells are remains an important subject for research.

Stem cells are the exception. They are, more or less, *unspecialized*; or, if you prefer, they are the specialists in specialization. There are different kinds of stem cells (see **Boxes 4.1** and **4.2**), a fact that has implications both for science and for politics, discussed below. They all, however, both reproduce themselves and produce other kinds of cells. (It is not unusual for embryonic stem cells in a petri dish to turn, quite spontaneously, into heart cells, possibly just because they want to turn into *something*.[5])

4.2 STEM CELLS BY ORIGIN

Stem cells can be classified according to where in the body, and when in its development, they are found. The two main categories are:

adult—multipotent stem cells that exist in every human, including children; in blood, bone marrow, brain and fat cells, among others
embryonic—technically, pluripotent stem cells existing in the first eight weeks after conception; often used to mean any non-adult stem cells

Embryonic stem cells are sometimes referred to as **ES cells**. A somewhat related category is that of **EG cells**, **embryonic germ cells**, which are also pluripotent and are very early developmental stages of the cells that make eggs and sperm. (Confusingly, EC cells are not a typo but something else entirely; they come from a form of cancer of the germ cells.)

Technically, an **embryo** exists for eight weeks after conception. At about five days, or one hundred cells, it becomes a **blastocyst**, and eventually the developing entity becomes a **fetus**. These distinctions are regularly blurred in media coverage.

Adult stem cells are sometimes, but not always, referred to by their location: as **brain stem cells**, **blood stem cells**, and so on. More generally, stem cells are often referred to as **progenitor cells**, because they are the "ancestors" of other kinds of cells. Lately that has become a useful euphemism, but it is also used to indicate some uncertainty as to a cell's true classification; some researchers, unsure if "brain stem cell" is an accurate term, call them "adult neural progenitor cells."

Other, somewhat more common, terms include:

fetal stem cell—pluripotent stem cells from the developing fetus
placental stem cell—likewise, from the placental tissue
umbilical stem cell—from the umbilical cord

These categories are more often used in political rather than scientific ways. The term "fetal" is sometimes used either to mean "derived from an abortion" or else simply "embryonic" (or both). In general, people with an ax to grind are liable to select the terms that evoke the emotions with which they hope to sway their audience.

Each of us began as one single cell—a fertilized egg—and developed into maybe 50 trillion cells of perhaps three hundred types.[6] *How* that happens, we don't exactly know, but stem cells, with their remarkable transformative abilities, may at the very least give us clues. One of the major arguments in favor of embryonic stem cell research is that it will enable this basic science to be studied.

If we could first understand and then control the process of cell growth, differentiation, and development, we might be able to grow tissues we could use for **regenerative medicine**. Theoretically that would let us regrow, say, the broken part of a spinal cord to let the paralyzed walk. But there is a long way to go before that is a realistic prospect.

■ SOURCES OF EMBRYONIC STEM CELLS ■

UNLIKE OTHER CELLS, embryonic stem cells, in the right conditions, keep on duplicating themselves. That's why there are often references to **stem cell lines.** Each line can give rise to many cells, but each line, of course, started somewhere.

The cells that form these lines are not easy to isolate; scientists may need to make many attempts (each with a different blastocyst) to isolate a single line. As it happens, many early embryos do exist, because fertility clinics (see **Chapter 5**) routinely create more than they proceed to implant. The remainder are typically frozen and stored, often for long periods. The only comprehensive survey on the numbers of embryos being stored was conducted in 2002, and found that there were 396,526 frozen embryos in the US.[7] Only a small proportion of those were officially "designated for research" (an estimated 11,000) but there had been little effort to encourage donation, let alone to demonstrate how immediately useful they might be.

These leftover embryos could be used as sources of stem cells. That does entail their destruction, so the action is controversial for at least some anti-abortion activists. On the other hand, the frozen embryos are most often destroyed anyway, in the end, so other conservatives are willing to consider their use. Senator Orrin Hatch, who is staunchly anti-abortion, has asked,[8] "Why shouldn't these embryos slated for destruction be used for the good of mankind?"

Another possible source of fetal tissue from which stem cells can be derived is aborted material. Both that and leftover embryos are generally considered ethically acceptable by most scientists (see **Box 4.3**), if not by many religious conservatives.

4-3 **SCIENTISTS' OPINIONS ON USING EMBRYOS**

A SURVEY, published in 2004, of members of the American Society of Human Genetics and its international counterparts, found strong but qualified support for research on embryonic stem cells, with Americans being somewhat more permissive than their international colleagues.[9]

QUESTION	YES %	
IS IT ACCEPTABLE TO ...	US	INT'L
use embryos retrieved from aborted fetal tissue?	77	63
use embryos that were created for fertility treatment?	59	51
destroy human embryos during research?	54	40
create human embryos specifically for research?	18	14

Clearly, even scientists working in the field have significant moral qualms about working with human embryos. Only 9 percent of American scientists (3 percent internationally) considered reproductive cloning acceptable, much the same proportion as in the general public (see **Chapter 9**); perhaps the distinction between "scientists" and "ordinary people" is less sharp than it is sometimes portrayed.

It is also possible to make an embryo specially—either by ordinary IVF techniques, but without any intention of creating a pregnancy, or by cloning. This is widely considered unacceptable among scientists (see Box 4.3), as well as the general public (see **Chapter 9**).

One further theoretical possibility is **parthenogenesis**. This is somewhat similar to cloning except that the egg is mysteriously stimulated to generate an embryo without either sperm or other tissue being used. Research has been done on this, as an end run around legal and ethical restrictions, but it seems unlikely to become an important source. (By definition, incidentally, the embryo created would be female, since no male can be involved, and any derived stem cells would not include a Y chromosome.)

▪ THE STEM CELLS EVERYONE LIKES ▪

RESEARCH ON **adult stem cells** is essentially uncontroversial. Unfortunately, the question of how useful they will be remains unresolved and a contentious issue among politicians and those who try to influence them. There have even been complaints within the scientific community that "those working on embryonic stem cells have turned and stabbed those working on adult stem cells in the back" as part of the political struggle.[10]

Adult stem cells are generally believed to be less malleable than embryonic stem cells—multipotent as opposed to pluripotent—but they do have one notable advantage: If they could be used for therapies, the patient could perhaps be the source for the curative cells. This would presumably eliminate problems of rejection, immune-compatibility, and so on, without elaborate work-arounds (such as cloning, discussed in detail below) or extensive drug treatments.

One by-now routine procedure, the bone marrow transplant, does in fact work because the marrow contains adult stem cells, albeit from a donor rather than from the patient: The transplanted stem cells proceed to rebuild the patient's blood and immune system. This became standard practice well before scientists knew exactly how it worked, and it is now becoming practical simply to transplant the stem cells themselves.

Several studies do indeed suggest that it may be possible to use adult stem cells for other purposes. For example:

- bone marrow stem cells from mice have formed "spine, kidney, skin, guts, brain and uterus" cells[11]
- they also seem to have reduced degeneration in the eyeballs of experimental mice that were designed [don't ask] to deteriorate[12]
- a patient's own blood stem cells seem to have helped repair his accidentally damaged heart[13]
- some mice with heart attacks have had their tissue repaired by adult stem cells from other mice[14]
- a patient's own brain stem cells, removed, cultured, and re-injected, seem to have reversed symptoms of Parkinson's[15]

These, and many other, examples represent genuine science, not yet fully understood or ready for practical use, but certainly promising. There have been criticisms of some of these findings, for example suggestions that the adult stem cells in these studies didn't really transform themselves into the needed tissue type, but rather fused with existing specialized cells; which, of course, merely indicates that more research is needed.

All these experiments are assiduously chronicled by those who are morally opposed to any use of embryonic stem cells (see below). Unfortunately for them, however, many of the scientists who produced these promising results actively support research on embryonic stem cells. Indeed, the publication, in *Nature*, of the first mouse bone marrow study in the list above was announced at a press conference that also, in a transparently political move, featured a report about embryonic stem cells. This is how the mouse researcher, Dr. Catherine Verfaillie, responded to a question about this:[16] "These are really inter-twined studies and they should go forward side by side."

To understand why this is a problem, we need to delve fur-ther into politics. And to understand the context of that, a lit-tle history is required.

■ A LITTLE HISTORY ■

SCIENTISTS HAVE BEEN working on the basic science of differ-entiation since the 1970s, mostly in mice (as usual), but without any expectation of near-term treatments, let alone cures. Prior to 2000, noted Dr. Stuart Newman, "no successful treatment of symptoms in an animal model of human disease [had] been achieved with embryo stem cells."[17] Moreover, "the distinguish-ing characteristic of mouse ES cells when they were first identi-fied was that they caused cancer when injected into mice."[18]

In fact, for many years it was difficult if not impossible to get grants to work on human embryonic stem cells, not so much because of ethical objections as because, according to Martin Evans, the man who first isolated mouse embryonic stem cells, "There was no impetus to do it. . . . We said, 'There is no point.'"[19]

In the 1990s, however, attitudes began to change. For vari-ous reasons, including the entrepreneurial efforts of a few sci-entists such as Michael West and William Haseltine (see below and **Chapter 10**), and the development of the Human Genome Project, interest in the feasibility of regenerative med-icine increased. Financed by Geron, which West had founded, several scientists worked on isolating embryonic stem cells, and on November 5, 1998, the University of Wisconsin put out a press release that began: "The dream of one day being able to grow in the laboratory an unlimited amount of human tissues for transplantation is one step closer to reality."

One step, but only one. Aside from the awkward fact that mouse stem cells tended to cause cancers—even if that could be avoided somehow—there remained the question, as with any transplant, of tissue matching.

Meanwhile, in Scotland, researchers had cloned a sheep. It was born on July 5, 1996 and announced the following February. This was the first mammal to be cloned. People, of course, are mammals, too, but the connection between cloning and stem cells was not at first obvious to most people. It soon would be.

■ BLUE SKY SCIENCE AND MINDLESS POLITICS ■

THIS IS THE POINT at which science moved into speculation. Very appealing speculation, and not completely unfounded, but definitely simplistic. Still, it made for a compelling story, which may first have been articulated by Michael West but rapidly spread around the world. This is how it went: "If you could clone yourself, not to make a baby but just to make an embryo, then if you could extract embryonic stem cells they would have your own DNA—so if you could use them to treat yourself there might be no rejection problem."

That sentence includes three uses of *if*, three of *could*, one of *would*, and a concluding *might*. As a way of getting attention (and funds), this is brilliant. It doesn't actually *promise* anything—so it's not wrong—but it outlines a dream of a cure that makes sense even (especially) to non-scientists.

But does the clone-yourself story make sense? The answer is almost certainly *No*, not in this simplistic form anyway. The two fundamental reasons are:

1. There could never be enough eggs to make it work for more than a tiny minority of those claimed as potential patients; see **Box 4.4**.
2. As a practical matter, customized medicine for one is not a reasonable approach; see **Box 4.5**.

4·4 HOW MANY EGGS WOULD BE NEEDED?

IN ORDER TO make a clone, you need an egg from which to remove the nucleus and into which you can insert a cell from the person to be cloned (see **Chapter 3**). In fact, you need many eggs, because the process is extremely inefficient. The Korean scientists who were the first to extract ES cells from a cloned embryo (see **Box 4.9**) needed 242 eggs.[20] Animal studies suggest this may be typical; researchers working with mice have required from 29 to 926 eggs to get one stem cell line.[21]

The trouble with human eggs is that they come from women, and not easily. A whole course of hormones, whose safety is in dispute, and a minor surgical procedure—never without risk—may lead to the recovery of fifteen eggs, possibly less. So any "treatment" following the cloning/stem cell pitch would almost certainly require many more egg donors than patients; in the Korean experience, sixteen donors for one set of stem cells (see **Box 4.9** again; the Korean news gets worse).

How many patients? The Coalition for the Advancement of Medical Research (CAMR), which prefers to say "SCNT" (Somatic Cell Nuclear Transfer) to avoid the connotations of "cloning," claims: "Nearly 100 million Americans suffer from cancer, Alzheimer's, diabetes, Parkinson's, spinal cord injuries, heart disease, ALS, and other devastating conditions for which treatments must still be found. SCNT could hold the key to ending these patients' suffering."

That's roughly one-third of Americans—let no one say CAMR thinks small—but perhaps they didn't really mean that each of those 100 million patients would require a personal clone. Another, somewhat more plausible, estimate is that 1.7 million therapies "could" be performed each year.[22] That would imply the need (at the Korean success rate) for 411 million eggs. Which in turn would require, at fifteen eggs per donor, some 25.7 million donors.

There are less than 30 million women in the United States between the ages of 20 and 34. (The top end of that age range is not ideal for egg donation, but let's be generous.) Some of them are pregnant, or wish to be, or have recently given birth, or have medical, ethical, religious, or other objections to donating eggs, or are simply too busy; at a complete guess, say half. That would leave 15 million available donors, when we need

4·4 25 million: Even if every available young woman donated eggs every single year, they couldn't keep up with the demand.

These numbers may be high. But even at a stunning success rate of fifteen eggs per stem cell line—that is, one donor per patient—1.7 million donors would be needed every year; almost as many women as turn twenty each year. In other words, either every woman donates eggs once or, more likely, some women sell them several times.

No matter how much you scale down the requirements, the "clone your own" story implies a scenario that turns young women into egg factories, and does so on an inconceivable scale. And which women? Poor women, of course. (Unlike the Yale undergraduates offered $50,000 for their eggs—see **Chapter 5**—these women's genes are not required, just the shell.)

This is simply not going to happen.

4·5 ## THE STORY THAT WILL NOT DIE

THE STORY ABOUT using stem cells from your own clone has been discredited many times. An April 2001 article in the prestigious magazine *Nature* opened:[23] "The idea of therapeutic cloning, which offers the potential of growing replacement tissues perfectly matched to their recipients, is falling from favour." The *New York Times* also debunked it, under the headline "Use of Cloning to Tailor Treatment Has Big Hurdles, Including Cost," quoting:[24]

> It's too laborious and costly to employ as a routine therapeutic procedure.
>
> —Dr. Alan Colman, PPL Therapeutics

> They're never going to have enough women's eggs available to do it.
>
> —Dr. Alan Trounson, Monash Institute of Reproduction and Development; advisor to ES Cell International

4-5

A few months later, the Los Angeles Times took the story apart yet again, quoting the CEO of Geron:[25] "Okarma said it would take 'thousands of [human] eggs on an assembly line' to produce a custom therapy for a single person. 'The process is a nonstarter, commercially,' he said."

Others agreed, including Alan Robins, chief scientific officer of BresaGen Ltd., a cell therapy company in Australia and Athens, Georgia:[26] "It is not something we want to get involved in. . . . We don't think it makes sense as a business model, producing cell therapies for a patient population of one."

In early 2004, when Korean scientists made an overhyped advance (see **Box 4.9**), the New York Times pointed out: "Despite Advance in Cloning, Scientists Are Tempering Hope With Reality."[27] The Los Angeles Times cautioned: "Clone Is One Step in Extended Process: Success with human embryo offers promise but stem cell treatment is a long way off."[28]

And so it continues to go. The story is just too appealing to die.

So why does the story survive? Some researchers definitely do seem convinced by it; the suspicion is that they are so focused on the interesting, and potentially very important, work they do that they have not considered the practical ramifications. Other scientists seem, at a minimum, willing to go along with it in order to promote technologies they believe will be worthwhile. Paul Berg, Nobel laureate and elder statesman, cowrote an article in April 2002 that spelled out this approach in detail, although only the month before he had dismissed concerns about a potential shortage of eggs with the claim that basic research on ES cells (from cloned embryos) would make the use of human eggs unnecessary.[29]

Whether it was used naively or cynically, the clone-yourself story did contribute very substantially to the hype about ES cells that arose in 2001, to be discussed shortly. But first, some political background.

■ THE POLITICAL EMBRYO ■

LESS THAN 20 percent of the American public think abortion should be illegal under all circumstances, even when the woman's life is threatened.[30] (A significantly larger number think it should be legal "only in a few circumstances.") Nevertheless, these "pro-lifers" have had, and do have, an enormous effect on research funding.

Their belief is that human life begins at the moment of conception and therefore any embryo, of any age, is a person who deserves protection. Clearly, any research or procedure that potentially damages an embryo—including the removal of stem cells—would be unacceptable to them. Indeed, some are disturbed by fertility treatments that involve generating more embryos than are implanted; there is a movement to "adopt" frozen embryos and bring them to term in a volunteer womb.[31] (Note, however, that many fail to implant or otherwise survive the process of pregnancy.)

Most scientists, as noted above, consider that **embryonic stem cells** hold even greater promise for medical treatments than adult ones, on the grounds that they are less specialized and therefore may more easily be induced to become whatever we want. (The adult-cell argument is that any individual patient would, presumably, only need one general kind of cell, so the multipotency of one particular kind of adult stem cell would be all that was needed.) But ESCs do come from embryos. And, like it or not, that puts us squarely in the realm of abortion politics.

As shown in **Box 4.3**, even scientists working in the field have significant qualms about working with human embryos, and especially with creating embryos specifically for research. Abortion opponents, naturally, take these concerns much further, to the extent that one characterized an attempt at compromise this way:[32] "Faced with a 99% death rate from cloning, such proposals would 'solve' the problem by ensuring that the death rate rises to 100%."

Frustrated scientists and their supporters, notably in the patients' rights community, tend to respond in rhetorical kind. As a result, the discussion of embryonic stem cells has become a ritualized sham that bears little relation to reality, in which both sides take absurdly extreme positions and defend them to the death of all reason. Caryl Rivers put it well in the *Boston Globe*:[33] "If some people fire off a shell with 'babykiller' on it, you lob one back that says 'grannykiller.' They say your side is chopping up babies, so you say their side wants millions of people who now have diabetes or Alzheimer's or other diseases to die painful deaths." That's not much of an exaggeration; see **Box 4.6**. Coming as it has from both scientists and religious conservatives, this rhetoric has confused the issue to the point where real questions remain not only unresolved but virtually unmentioned.

4.6

"YOU ALWAYS GO WITH THE LITTLE GIRL."

PERHAPS THE MOST extreme public manifestation of emotional arguments on both sides came on July 17, 2001, when Congress was considering the funding of research into embryonic stem cells. Either giving testimony before a subcommittee or addressing a press conference outside, were:

- a man holding two babies: "Which of my children would you kill?"[34]
- a twelve-year-old girl: "All [my sister] wants to do is live a normal, healthy life, and embryonic stem cell research is our best hope."[35]
- a mother describing her two-year-old daughter, "an ambassador for the roughly 188,000 frozen human embryos like her . . . who could be adopted rather than terminated."[36]
- a woman with Lou Gehrig's disease: "You have the choice to be pro-life for an unimplanted frozen embryo that will be discarded or pro-life for me. . . . I am asking you to choose me."[37]

No one ever wanted to kill that man's babies. No twelve-year-old is competent to compare the prospects of different research protocols. No two-year-old is any kind of ambassador. And, rough as it may seem to say

4.6 so, by the time any cures could be developed, on the most optimistic scenario, the last witness will probably have succumbed to her disease.

But advocates think this kind of testimony works. Dr. Robert Goldstein, the Chief Medical Officer of the Juvenile Diabetes Research Foundation, was asked in 2004 if he'd send scientists to Congress or a young patient, "Are you kidding?" he asked. "You always go with the little girl. There's no choice."

This is no way to make policy.

To their credit, some stalwarts of those culture wars, on both sides, have tried to get away from the knee-jerk responses and unthinking hostility of the abortion morass. They include the staunchly anti-abortion Senator Orrin Hatch, and the equally fervently pro-choice Judy Norsigian of *Our Bodies, Ourselves*—who actually ended up disagreeing again, this time about cloning for research, which he supported and she opposed (see **Chapter 11**).[38] (Of her relationship with Hatch's usual allies, Norsigian wryly remarked, "This may be the only issue on the face of the Earth we agree on.") Unfortunately, they have had only limited success in bringing new perspectives to the issue.

■ FEDERAL FUNDING OF EMBRYO RESEARCH ■

THE IMMEDIATE PRACTICAL consequence of the mixing up of abortion politics and basic science is that federal funding of research involving embryos—any research, not just that on stem cells—dried up long ago. (A very vocal minority can always have a disproportionate impact.) Several different ways of preventing it were used, including the strictly bureaucratic— requiring the approval of an ethics board that did not exist—culminating in the Dickey-Wicker Amendment to the federal budget of 1996. Renewed annually since, this prohibits funding of work,[39] "in which a human embryo or embryos are destroyed, discarded, or knowingly subjected to risk of injury or death greater than that allowed for research on fetuses in utero."

Researchers, by large margins (71 percent in one poll), think this prohibition should be lifted; and by an even larger one (87 percent) think that it does not apply to embryonic stem cells, which cannot by themselves develop into a fetus.[40] (By implication, at least, President George W. Bush agrees; see below.)

One final, almost Machiavellian, reminder: As noted at the end of **Chapter 3**, some advocates of embryonic stem cell research, and Human GE in general, seem to be relatively satisfied with this state of affairs. The Biotechnology Industry Organization (BIO) has been criticized for unconstructive lobbying, and activists such as Dr. David Gollaher, Dr. Paul Berg, and others prefer no law to what they call a "bad law."[41] Stalemate works for them—precisely because it does prevent meaningful regulation, and they are therefore not prohibited by law from carrying out experiments, just so long as they can find the money.

■ THE BUSH COMPROMISE ■

IN 2001, the then newly elected George W. Bush came under heavy and increasing pressure to allow federal funding of embryonic stem cell research. This was partly because of an active campaign in favor of it, which to some extent arose because of fears among researchers and patients' advocates that Bush would favor shutting down such research altogether. That in turn may be why the clone-your-own story gained so much traction at that time.

A publicly committed opponent of abortion, Bush was also under strong pressure *not* to allow funding of work that would involve the destruction of embryos. He was being urged by his natural constituency to support a complete ban on all forms of cloning, including those that were not intended to lead to pregnancy, and to do what he could to prevent any kind of embryo research.

To the surprise of almost everybody, Bush found a compromise. On August 9, 2001, in the first nationally televised speech of his presidency, he announced that federal funding *would* be available. He did not say how much, though he specified that $250 million would be spent on "research on umbilical cord placenta, adult and animal stem cells which do not involve the same moral dilemma."[42]

The catch was that federal funds would be available *only* for work on stem cell lines that then existed (**Box 4.7**). No new embryos could be destroyed; but lines that had already been generated through the destruction of embryos could be used.

4.7

WHICH LINES ARE ELIGIBLE FOR FEDERAL FUNDING?

PRESIDENT BUSH, in his August 2001 speech, referred to "over sixty" lines of embryonic stem cells. Before that, only seven had been published in the scientific literature.[43] On August 27, 2001, the National Institutes of Health (NIH) identified sixty-four lines, later increased to seventy-eight. That turned out to be extremely overoptimistic.

On March 1, 2005, the NIH website listed twenty-two lines "available for shipping" (up from fifteen a year earlier) that were eligible for federal funding, and forty-four (down from fifty-one) "not yet available for shipping."[44] Four other lines had been withdrawn or had "failed to expand into undifferentiated cell lines."

Some of the "available" lines are proprietary and expensive. Others come from foreign countries, which may impose restrictions, as the Indian government has done. (Ten of the sixty-four lines were Indian; it is not clear why the government stepped in, unless it was to protect a potential growth industry.) Other countries with lines on the list include Sweden, Singapore, Korea, and Israel.

Finally, it seems that all, or at least most, of those lines are contaminated with mouse molecules. They may be useful for research, but not for therapies—though some scientists think they can be purified.[45]

Bush's speech served temporarily to defuse the issue. Some of his anti-abortion supporters were still unhappy, saying for example that "President Bush broke his word to the American people," but others followed the lead of the Rev. Jerry Falwell, who said, "I can live with it."[46]

■ THE PRIVATIZATION OF RESEARCH ■

BY FEBRUARY 2004, the National Institutes of Health (NIH) had awarded $60 million in grants for human ES cell research, but this was much less than was available. Dr. James Battey, chair of the relevant committee, explained that "a lack of applications, rather than NIH restrictions, is the reason the federal government has funded comparatively few stem cell grants."[47]

Many scientists just gave up on the government. Not only did they consider the number of ES cell lines inadequate, but the climate of regulatory uncertainty and executive displeasure was enough to make them vote with their feet—literally, in the case of Professor Roger Pedersen, who moved from the University of California at San Francisco (UCSF) to Cambridge, England, in order to continue his work on stem cells.

Those who stay in the US but choose not to accept federal funding find that research life gets much more complicated. The logistical and financial difficulties to be overcome are enormous, since all work has to be done in facilities that are *in no part* financed by federal funds. One response is to set up new, separate research centers, which can contain the money and avoid any possibility of illegality. Harvard is building a $100 million stem cell center; Stanford received a $12 million donation and expects more from California's increased state funding in 2005–14; as does UCSF, which had already raised at least $5 million.[48]

The other alternative is to work strictly in the private sector, but even there, as ACT's Dr. Robert Lanza admitted in early

2004,[49] "All the money for this work has dried up. We are lucky to still be in business. Our research has suffered immensely."

Several private companies, however, are working on adult stem cells (see **Box 4.8**). Fewer US companies were pursuing embryonic stem cell research in 2004. Geron had changed direction and laid off most of its stem cells staff, though it appeared to be keeping its options open for expansion in 2005 and subsequent years. ACT and VistaGen Therapeutics also remained committed to this line of work.

COMPANIES WORKING ON ADULT STEM CELLS

BIOTECH COMPANIES are often unstable and tend to be secretive, but among those thought to be pursuing research on adult—but probably not embryonic—stem cells in the US are:

- Geron (www.geron.com)
- Amgen (www.amgen.com)
- Osiris (www.osiristx.com)
- StemCells (www.stemcellsinc.com)
- Cambrex (www.cambrex.com)
- Cytokinetics (www.cytokinetics.com)

Others that used to be in the field include Advanced Tissue Sciences Inc. (www.advancedtissue.com), NeuralStem Biopharmaceuticals (neuralstem.com), Layton Bioscience (laytonbio.com), SyStemix Inc., and Celltrans Inc.

Curis Inc. (www.curis.com) used to work on adult stem cells but transferred its patent rights on December 18, 2002 to the Singapore-based company ES Cell International Pte Ltd (ESI), which also works on embryonic stem cells, and supplies several of the "legal" lines on the NIH register.

The Korean research that led to the somewhat overhyped 2004 announcement of stem cells from a clone (see **Box 4.9**) would have been legal in the US, since it was privately funded.[50]

That same research has, however, been described by a leading Korean citizens rights activist as having "moved beyond the national consensus."[51] It was suspended shortly after it made international headlines; but new government guidelines let it resume in 2005. The lesson seems to be more that similar debates are occuring in every country where the research is feasible than that the US is uniquely hampered.

4-9 DECONSTRUCTING THE KOREAN "CLONE FOR STEM CELLS"

IN FEBRUARY 2004, Korean scientists reported that they had harvested embryonic stem cells from a clone. As is common, their achievement was badly overstated in much of the media.

The scientists commented, following the standard script:[52] "Because these cells carry the nuclear genome of the individual, after differentiation they could be expected to be transplanted without immune rejection for treatment of degenerative disorders. Our approach opens the door for the use of these specially developed cells in transplantation medicine."

Up to a point. In fact, the experiment only worked when the person being cloned was the egg donor. And then only when the cells used for cloning were cumulus cells, which "are found in the ovaries and in some species have been found to work especially well in cloning experiments."[53] The researchers also said, quite seriously, that "our Korean finger techniques helped. Koreans eat with metal chopsticks, which are very slippery."[54] In other words, the process worked only for one of the groups least likely to need it: healthy, fertile, young women—who live in Korea.

The experiment did have value, unlike a similar one that ACT used in 2001 to get a lot of headlines. (Their claim to have cloned a human embryo—also using cumulus cells—turned out to describe an almost entirely unsuccessful attempt, which one expert called "a far cry from something useful."[55]) But the value is much closer to basic science than to the promised therapy. Professor Roger Pedersen commented,[56] "This will likely accelerate the development of alternative ways of reprogramming human cells, which could in the future diminish the need to use human eggs for this purpose."

4-9 Perhaps the most astringent comment came from Hilary Rose, a British academic with a long interest in these issues:[57]

> Here we go again. Reading the excited claims for the medical benefits likely to accrue from the Korean veterinary researchers' success in growing cloned human pre-embryos, one is entitled to feeling a certain déjà vu. Heading the list were those old favorites, treatments for Parkinson's and Alzheimer's disease. There really needs to be a phrase to describe this researchers' equivalent of the old charge against doctors of shroud waving.

Some other countries are moving forward, notably Singapore and the UK, which has opened a "national stem cell bank" to encourage research on embryonic stem cells. This remains controversial, although the UK does have one of the more stringent systems of regulation (see **Chapter 12**).

In practice, and in retrospect, the Bush compromise was hardly a compromise at all. It was a way of avoiding a decision about how to regulate this research. It reinforced the stalemate in Washington, and one consequence was that the action moved to the states, where the biotechnology industry has been putting its muscle and money behind attempts to write legislation on a state-by-state basis that will explicitly allow embryonic research and, preferably, fund it (see **Chapter 12**). California led the way with legislation in 2002, and $300 million a year allocated in 2004, with New Jersey, Massachusetts, and New York following.

■ WHAT'S NEXT? ■

THE RAMIFICATIONS OF the 2004 elections for stem cell research, both nationally and at a state level, particularly in California, will not be clear for a long time. As David Magnus and Arthur Caplan put it, in a December 2004 opinion piece,[58]

"The embryonic-stem-cell debate is dead. Long live the stem-cell debate."

The reelection of President Bush presumably means that federal funding restrictions will not be lifted. But California funding will increase dramatically and it may be that private funds will also help to fill the gap, now that it seems likely that the gap will persist. There may indeed be enough funds, though not necessarily well-placed or well-supervised (see **Chapter 12** for much more discussion of this).

Many questions remain, about scientific efficacy, about fairness, about profits and conflicts of interest, about informed consent (especially from the desperately sick), and indeed about the level of predictability that would make human trials ethically acceptable. As Magnus and Caplan conclude, "Now the debate about stem-cell research must focus on making sure that this research is done following the highest ethical standards."

■ FURTHER READING ■

Free Documents from the Web

The Council for Responsible Genetics (CRG) *Stem Cell Primer* and *Stem Cell Myths* are short (3 pages each, about 36k as pdfs) summaries of the issue, available from www.gene-watch.org/programs/cloning.html, where there are also links to several articles on related issues.

The Center for Genetics and Society (CGS) research cloning discussion at www.genetics-and-society.org/technologies/cloning/research arguments.html is unusually balanced and details arguments both in favor of and against research cloning, with rebuttals to each and many links, concluding by supporting a moratorium on the practice.

The National Institutes of Health (NIH) Stem Cell Information site at http://stemcells.nih.gov includes detailed information for researchers, and also a "Stem Cell Basics" page, available as a cumbersome eight-page pdf (960k).

The President's Council on Bioethics, *Monitoring Stem Cell Research,* is a 433-page report, available as a single pdf (1.5mb) or a series of web pages, from www.bioethics.gov/reports/stemcell/index.html. An excellent, if dry, overview of the field as of January 2004, it does not make recommendations but does include ten commissioned papers by experts in the field. The only trouble is, if you can wade through it all, you probably don't need to.

Books

Brian Alexander, *Rapture: How Biotech Became the New Religion,* Basic Books, 2003, is an enjoyable read that includes much on stem cells and revealing portraits of Michael West and others.

Stephen S. Hall, *Merchants of Immortality: Chasing the Dream of Human Life Extension,* Houghton Mifflin Company, 2003, is a deeper, though still accessible, book that also includes much on stem cells and revealing portraits of Michael West and others.

■ ENDNOTES ■

1 Cibelli and Rothstein quoted in Jonathan Bor, "Stem Cells: A Long Road Ahead," *Baltimore Sun,* 03/08/04; Andrews quoted (brackets in original) in Brian Alexander, *Rapture: How Biotech Became the New Religion,* Basic Books, New York, 2003, p.204

2 Bernadette Tansey, "Stem Cell Initiative Aids State," *San Francisco Chronicle,* 11/04/04

3 Judy Norsigian, Executive Director, Our Bodies Ourselves, 03/05/02.

4 Peter Mombaerts, "Therapeutic Cloning in the Mouse," *Proceedings of the National Academy of Sciences,* 09/20/03

5 Alexander, *Rapture,* op. cit., p. 201

6 How many cells a human has is a rather silly question, since bigger people have more, but estimates run from 10 to 100 trillion. The "300 cell types" statistic is from a *Los Angeles Times* article (05/10/02); other sources say 220 or less or more.

7 RAND Research Brief, RB-9038 (2003), summarizing: Hoffman, D.I., et al., "Cryopreserved Embryos in the United States and Their Availability for Research," *Fertility and Sterility* 79 (5): 1063–1069, 05/03

8 Anuj Gupta, "The Personal Sides of the Stem Cell Debate," *Los Angeles Times,* 07/18/01

9 Survey conducted by Isaac Rabino, PhD., reported in *Genetic Engineering News,* Vol. 24, No. 4, 02/15/04

10 Dr. Ammon Peck, University of Florida, quoted in the *National Journal,* 04/19/03

11 *San Francisco Chronicle,* 06/20/02

12 Otani, A. et al, "Bone marrow-derived stem cells target retinal astrocytes and can promote or inhibit retinal agniogenesis," *Nature Medicine*, 09/02

13 *Wired*, 03/07/03

14 "Scientists Repair Damage From Heart Attack Using Adult Bone Marrow Stem Cells in Mice," NIH summary of a paper in *Nature*, 04/05/01

15 *Toronto Star*, 04/09/02; *Washington Post*, 04/09/02

16 *San Francisco Chronicle*, 06/20/02

17 Posting by Dr. Stuart A. Newman, Professor of Cell Biology and Anatomy, New York Medical College, to an e-mail list, 08/04/01

18 ibid., citing Martin, G. R., "Isolation of a pluripotent cell line from early mouse embryos cultured in medium conditioned by teratocarcinoma stem cells," *Proc Natl Acad Sci U S A* 78(12):7634–8, 12/81

19 Quoted in Alexander, *Rapture*, op.cit., p. 115

20 AP 02/12/04

21 *New York Times*, 12/18/01

22 Cited by Andrew Kimbrell in testimony before the Senate Judiciary Committee on 02/05/02

23 *Nature*, 04/05/01

24 *New York Times*, 12/18/01

25 *Los Angeles Times*, 05/10/02

26 ibid.

27 *New York Times*, 02/15/04

28 *Los Angeles Times*, 02/13/04

29 Paul Berg, J. Michael Bishop, Andrew S. Grove, "Break Impasse Over Stem-Cell Therapies," *San Francisco Chronicle*, 04/17/02; personal interview, Sacramento, 03/20/02

30 Gallup conducted twenty polls about abortion between September 1991 and August 2001; on average, 15.25 percent opposed abortion under all circumstances, with the percentage varying from 12 to 19. Figures collated at www.frtl.org.

31 See www.snowflake.org, so named because "like snowflakes, each embryo is fragile, unique and the most beautiful of God's creations."

32 Richard M. Doerflinger, testimony to the Senate Commerce Subcommittee on Science, Technology and Space, 05/02/01

33 *Boston Globe*, 08/18/01

34 www.CNN.com, 07/18/01

35 *Boston Globe*, 07/18/01

36 *Los Angeles Times*, 07/18/01

37 Reuters, 07/18/01

38 *New York Times*, 06/18/01; *San Francisco Chronicle*, 08/09/01

39 Quoted in Stephen S. Hall, *Merchants of Immortality*, op. cit., p. 119; see pp. 98–122 for more details.

40 Rabino, I., "Geneticists' Views on Embryonic Stem Cells," *Science*, 293:1433–1434, 2001; cited in Rabino, *Genetic Engineering News*

41 Asif Ismail, "Lobbying, Old-Time Politics Block Legislation on Human Cloning," Center for Public Integrity, 02/15/01. The "bad law" quote is from a Stanford, CA, panel discussion, 03/10/02.

42 President G. W. Bush, "Remarks by the President on Stem Cell Research," 08/09/01

43 Hall, *Merchants of Immortality*, op. cit., p. 263

44 www.stemcells.nih.gov/registry; only current numbers are posted.

45 Karen Kaplan, "Study Says All Stem Cell Lines Tainted," *Los Angeles Times*, 01/24/05

46 *San Francisco Chronicle*, 08/11/01

47 AP, 02/13/04

48 *Boston Globe*, 02/29/04; *San Francisco Chronicle*, 08/19/02

49 AP, 02/13/04

50 *Boston Globe*, 02/13/04

51 *Nature*, 05/06/04

52 Quote taken from a press release issued by the American Association for the Advancement of Science, 02/12/04

53 London *Guardian*, 02/12/04

54 Interview with Dr. Woo Suk Hwang (speaking) and Dr. Shin Yong Moon, *New York Times*, 02/17/04

55 The *Economist*, 12/01/01

56 London *Independent*, 02/13/04

57 London *Guardian*, 02/16/04

58 David Magnus and Arthur Caplan, "Stem-Cell Research Will Now Proceed; The Issue Is How," *San Jose Mercury News*, 12/13/04

5

SELLING HUMAN GE

■ INTRODUCTION ■

N O ONE IS going to force you to to have designer babies. Not exactly. What they hope to do is *sell* them to you. Gregory Stock (see **Chapter 10**) predicted in his 2002 book, *Redesigning Humans*,[1] "With a little marketing by IVF clinics, traditional reproduction may begin to seem antiquated, if not downright irresponsible. One day, people may view sex as essentially recreational, and conception as something best done in the laboratory."

This is not a joke. Stock may be playful, and deliberately provocative, but he is also absolutely serious. He presents what he calls "Germinal Choice Technologies" as consumer issues, which is the common line among advocates of Human GE. And what can be bought will of course be sold (see **Box 5.1**).

Forget about mad dictators, cloned armies, and superhuman spies; think instead about advertising and carefully directed peer pressure. It's not the government that's going to promote human genetic engineering, it's commercial interests. What they want to do is keep the government out of it.

Why do advocates of Human GE think this is plausible? Because it's only one step on from what's happening now, particularly in the US. We already have:

- ♦ routine ads for "egg donors" (with cash almost always in the headline)
- ♦ TV ads for a breast cancer test that 99 percent of women don't need[2]
- ♦ a reality TV show about multiple cosmetic surgeries[3]

- a patient using a billboard to solicit a liver donation[4]
- sex selection being advertised in the *New York Times*[5]

5.1 "CAN YOU MAKE MY KID SMARTER?"

LEE SILVER (see **Chapter 10**) wrote this speculative piece, which was published in *Time*, about a fertility clinic in 2025:[6]

> The St. Gen spokeswoman [said] that anyone could call the technology whatever they wanted, but Organic Enhancement was the term St. Genevieve had chosen to use. "This is entirely appropriate," she said with a straight face, "since the DNA molecules added to embryos are totally organic."

> What will these "organic" enhancements do? According to this fanciful prospectus, they will: ". . . provide your child with all-natural resistance to heart disease, hypertension, diabetes, stroke and eight different forms of cancer, as well as absolute protection against AIDS, allergies, asthma and Alzheimer's disease." Note how every one of these selling points is concerned with avoiding disease (and, as written, probably oversold at best). But the title of the piece is "Can You Make My Kid Smarter?" and it ends: "Meanwhile, the scientists at St. Gen had their eyes on the future. A mere thousand genetic changes had been identified that were mostly responsible for the difference between the intelligence of chimpanzees and humans. Now if they could just ratchet up those genes . . ."
>
> . . . they could make a fortune. That's the message. So marketing that begins by urging prospective parents to give their children a better chance of resisting disease slides inexorably toward offering them a chance of higher intelligence.

Some of what's happening would be illegal in many countries, notably the sale of human sperm and eggs, which is in fact illegal in Louisiana, though not other states (see **Box 5.2**). Legal confusion is quite normal in the whole area of assisted reproduction technology (ART; see **Box 5.3**). Contracts for commercial surrogacy, for example, in which one woman is paid to bear another woman's child, are legal in some states, illegal in others, and not specifically covered by the law in most.

5.2 GOT EGGS?[7]

THE SALE OF women's eggs may be the most intimate example of commodification. Perhaps numbed by familiarity with the concept, some stem-cell advocates seem to assume that they can buy as many eggs as they want or need (see **Chapter 4**), ignoring both practical limitations and the potential for harm to the seller.

Some rich individuals take a more discriminating approach: There have been reports of Ivy League students being offered as much as $50,000 for their eggs, if they are intelligent, athletic, at least five-foot-ten and with a 1400 score on the SATs.[8] (A society in which healthy, intelligent, tall, good-looking young women with an adequate secondary education might be tempted to sell body parts would seem to have problems worth discussing and fixing.)

Informed consent is a serious issue with these procedures, particularly since the risk factors are not entirely known.[9] True believers in market mechanisms, however, see nothing wrong with such sales in principle. It may be significant, however, that even they refer to sellers as "donors." If something is too repulsive to be called by its proper name, should it not be eliminated?

5.3 ACRONYMS RELEVANT TO THE FERTILITY INDUSTRY

Acronym	Spelled Out	Notes
ART	Assisted Reproduction Technology	The general term
IVF	In Vitro Fertilization	Generally, the technique or practice of fertilizing an egg in the laboratory; the embryo is then normally transferred to the woman's uterus

5-3

Acronym	Spelled Out	Notes
GIFT	Gamete Intrafallopian Transfer	A variation of IVF in which fertilization takes place in the fallopian tube
ZIFT	Zygote Intrafallopian Transfer	A variation of IVF, in which the embryo is transferred to the fallopian tube
ICSI	Intracytoplasmic Sperm Injection	A method of fertilizing an egg that requires very little sperm
PGD	Preimplantation Genetic Diagnosis	Used to check the genetic makeup of a very early embryo
ASRM	The American Society for Reproductive Medicine	Formerly the American Fertility Society (founded in 1944); the leading industry organization, publisher of Fertility and Sterility, an academic journal
SART	The Society for Assisted Reproductive Technology	An organization of fertility doctors, with 370 members representing 95% of US ART clinics
GCT	Germinal Choice Technologies	Invented by Gregory Stock to describe the combination of ART and Human GE; rarely seen outside his writings

For more about the technical procedures, see **Chapter 2**.

Basically, the market rules. One of the leading entrepreneurs in the area refers to his operation, the Reproductive Genetics Institute of Chicago, as being "at the core of capitalism."[10] He's right. In fact, for better and worse, that is how the fertility industry has always been, in the US.

The noted legal expert Lori Andrews has pointed out,[11] "In vitro fertilization itself was applied to women years before it was applied to baboons, chimpanzees, or rhesus monkeys, leading some embryologists to observe that it seemed as if women had served as the model for the nonhuman primates."

Fertility treatments have always been treated differently from other medical procedures, and there is one very obvious reason why (as well as a few less obvious ones, mentioned below): The patients—or some of them—are willing, indeed eager, to try almost anything. They just want a baby.

■ THE FERTILITY INDUSTRY ■

LOTS OF PEOPLE can't have babies without medical help. Roughly 10–15 percent of straight couples trying to have a baby do not succeed within a year; that's the standard rule of thumb for identifying infertility. It's a bit misleading; many "infertile" couples eventually have children without treatment; many more succeed by taking medication or having relatively routine surgery. Less than 5 percent of patients—less than 1 percent of people—need IVF and similar procedures.

5.4

THE EXPANDING INDUSTRY OF REPRODUCTION

THE FIRST "test-tube baby" was born on July 25, 1978, and with her the modern fertility industry. Within twenty-five years, it had become an estimated $4 billion business in the US. Between 1996 and 2001 alone, the number of babies born as a result of ART treatment doubled:[12]

	1996	2001
ART CYCLES[*]	64,724	107,587
SUCCESSFUL PREGNANCIES	14,573	29,344
BABIES BORN	20,921	40,687[†]

* Each cycle of treatment consists of several procedures for one woman, some of whom probably went through more than one cycle (data is not recorded by individual).

† In 2001, 38.4 percent of successful ART pregnancies resulted in twins, 3.3 percent in triplets or more. The rates in 1996 and 2000 were similar.

Nationally, there were 4,026,036 births in 2001, so ART accounted for almost exactly 1 percent of babies born.

5.5 THE COST OF ART SERVICES IN THE US

ONE "PROMINENT CLINIC" IN THE US IS QUOTED AS CHARGING:[13]

Initial consultation	$370
One IVF cycle using never-frozen embryos	$9,345
Transfer of frozen embryos	$4,000
PGD (for sex selection or disease screening)	$4,000
ICSI (frequently a prerequisite for PGD)	$2,000
Preconception sex selection (e.g. sperm sorting)	$2,000

Not all these charges may be incurred, but two or more cycles of treatment may be needed, with the second perhaps being a transfer of frozen embryos. Surrogacy or the purchase of eggs or sperm would be extra, as is the cost of medical care during and after pregnancy.

Some clinics offer discounts, sometimes in exchange for egg "donation," some with a partial money back guarantee. Prices do vary, but these are not atypical; see also **Box 5.6**.

Those who do need the cutting-edge treatments, however, absolutely depend on them. And the triumph of the industry (which is rapidly growing—see **Box 5.4**) is that it has developed techniques that, separately or in combination, can help almost anyone. (The most important are discussed in **Chapter 2**.) Its shame is that they are so expensive that most people cannot afford them (see **Box 5.5**).

The experience of the patients—or customers—served by fertility clinics is hard to characterize accurately and sensitively. They pay a lot of money and, if all goes well, receive a literally priceless good. Apologists for the present, largely unregulated, system (or lack of system) rely on this as an argument for keeping things the way they are. The evidence from Europe, however, suggests that customers—or patients—in the US are, on balance, being badly exploited (see **Box 5.6**).

5.6 COMPARISON OF US WITH EUROPEAN ART COSTS

HEALTHCARE IS more expensive in the US than in Europe, and ART is no exception, as these 2001 numbers show:[14]

	US	EUROPE
COST PER IVF CYCLE	$9,226	$3,531
COST PER LIVE BIRTH	$56,419	$20,522

Another cost factor is the incidence of **multiple births**. Multiples are much more common in the US than in Europe:

	US	EUROPE
INCIDENCE OF TWINS	32%	26%
INCIDENCE OF TRIPLETS OR MORE	7%	2%
MULTIPLES AS % OF BABIES BORN	57%	43%

These are 1998 figures, but the incidence of multiples in the US has not dropped significantly in recent years. The discrepancy may be the result of commercial considerations—any birth is a success for the ART clinic, even if multiple births are problematic for the parents.

Hospital charges for twins average four times those of singletons, while for triplets they reach eleven times, over $100,000. That's just for the birth. Premature and low-weight babies, as multiples tend to be, are in general more likely to suffer from chronic, and therefore expensive, medical conditions, including "blindness, respiratory dysfunction and brain damage."[15] There are, however, no definitive follow-up studies on the health of ART multiples, or indeed on ART children in general.[16]

The difference to the pocketbook of the infertile consumer is even greater than the dollar figures would suggest, because much of Europe provides public funding for ART (and for healthcare in general). Most developed countries recognize infertility as a medical condition. For most Americans, however, it is not even covered by insurance. Not surprisingly, therefore, per capita use of ART is much higher in most European countries—as much as five times the US rate in Denmark, Finland, and Iceland.

5.6 The Europeans are not technologically behind the US in this field, and never have been—IVF was pioneered in Britain, and ICSI in Belgium. They have just made different social decisions.

The argument from (limited) success is not the only reason for the lack of systematic regulation in this area. There is considerable reluctance on all sides to engage in what could be yet another front of the abortion rights struggle. Those who are unalterably opposed to abortion tend to have qualms, at least, about the fertility industry's practices, but in general seem to regard it as a low priority issue. Abortion rights supporters are understandably reluctant to see legal regulations applied to essentially medical decisions.

For example, more than one embryo is typically transferred (often three), since there is a notable failure rate in the process.[17] At one time, it was common to transfer several embryos—six, seven, or more—and then, if necessary, perform a "selective reduction." This is generally no longer seen as effective or necessary, but the optimum number remains controversial and is at least partly a judgment call: Multiple transplantation may increase the chances of a live birth, but multiple births are dangerous and expensive.

And the insurance industry clearly does not want to pay for an increase in medical services. Put all this together, and inertia becomes very difficult to overcome. The proponents of Human GE are counting on that. Ironically, they might actually be the ones to provoke reform.

■ "GOD DOESN'T MAKE BABIES—I DO"[18] ■

MOST DOCTORS specializing in obstetrics and gynecology (ob-gyn) are undoubtedly conscientious physicians. They are in general well-compensated (average incomes of $233,061), but they make less than surgeons or anesthiologists and they are highly trained and performing a useful service.

There are of course exceptions. Occasional errors are to be expected; the professional associations—and the law, when it comes to that—condemn abuses and sanction offenders. A few of those in the fertility industry, however, demonstrate attitudes that are disquieting, to say the least, even when their actions are not illegal and perhaps not even contrary to their code of professional ethics. As one patient's attorney said:[19] "The whole thing is creepy . . . I kept returning to the feeling that this area of medicine is unregulated and these in vitro doctors are like gods. They have complete power over these desperate people who want to become pregnant."

Another doctor—the one quoted in the heading of this section—routinely used *his own* sperm, instead of sperm from anonymous donors or, at least once, instead of the husband's. He seems to have believed he was doing the women a favor. After all, who wouldn't want their baby to have the genes of a successful medical practitioner?

(The doctor in question was discovered, it seems, because at least one baby looked like him. DNA tests proved that fifteen, and perhaps as many as seventy-five, of his patients' children were from his sperm. He had the nerve to claim that he was safeguarding patients from AIDS, and appealed his conviction all the way to the Supreme Court. The crime? Officially, mail fraud and wire fraud, which, as Lori Andrews wrote, "was sort of like getting Al Capone on tax evasion."[20] He got five years, a $75,000 fine, and was ordered to refund $39,205.)

Yet other doctors show remarkable insensitivity to their patients, for example over the issue of multiple births. As discussed above, there is a serious judgment call about how many embryos to transfer. So, are multiple births a success or a failure? Said one practitioner, dismissively, "Certainly they're a success for the couple."[21] (It would be most accurate to say that they are a success for the *clinic*, which reports births as a percentage of attempts.) An even more brutal response is the one given—*by their doctor*—directly to mothers who are worried about having triplets or quads,[22] "What do you mean you're not happy? This is everything you wanted and then some."

To repeat, it's the *attitude* that is the issue here. One of the people most prominently claiming to attempt human cloning, Dr. Zavos (see **Chapter 3**) encapsulated it, perhaps unwittingly, during a 2002 presentation. (Zavos is *not* an ob-gyn, but his wife is and together they run a fertility clinic.) He showed a slide of a needle poking into an egg and commented,[23] ". . . and then you rape the egg with the needle."

■ EXPANDING THE MARKET ■

MODERN ASSISTED-REPRODUCTION technology was developed to treat people who could not otherwise have children. But think how much larger the market would be if the industry could sell its services to people who are *not* infertile—which is most of us (85–90 percent). If only one in twenty of the fertile were to participate, that would more than triple the number of clients.

Some clinics have already started to appeal to the broader audience. There are several ways in which this is happening, some of which are directly relevant to Human GE, while others at a minimum commodify reproduction and exacerbate the trend of turning a natural process into a salable commodity.

For example, some clinics, in the US and abroad, are trying to sell the "convenience" of **egg freezing** to career-oriented young women. The idea is that a woman's eggs deteriorate over time—hence, in part, the "biological clock"—but, as one such company points out, "20% of women wait to begin their families until after age 35."[24] The techno-fix: Freeze your own eggs in your mid-twenties and use them in your late thirties—or even later. The world's record for giving birth is thought to be sixty-six years old.[25] The American record is sixty-three, by a woman who told her doctor she was fifty, which authorities recommend as the maximum age for ART, for many obvious reasons.

According to the official statistics, success rates with frozen donor *embryos* are barely half those with fresh ones (27.3 births per transfer, compared with 47.0).[26] As noted in Chapter 2, frozen *eggs* are such a new concept that there is little reliable data on success rates, which seem to be even lower.[27] In fact, the company quoted above is careful not to offer a guarantee.[28] Moreover, there remains considerable concern about the medical consequences of taking fertility drugs to hyperstimulate the ovaries in order to produce the eggs used for (in this case) freezing and storage.[29] Not to mention the financial costs: The most expensive services can run to $14,000, and some women choose to have two or three treatments. And then there are possible physical and psychological issues for both mother and child for late motherhood . . . If a human society is set up so as to be inconvenient for humans, would it not make more sense to adjust the society?

■ SEX SELECTION ■

THE EASIEST "SERVICE" to understand, and in some ways both a link to the benighted past and a harbinger of a potentially frightful future, is the choice as to whether a child will be a boy or a girl. Most ethicists, as well as most of the general public, view this idea with distaste, but according to *Fortune*,[30] "In surveys, a consistent 25% to 35% of parents and prospective parents say they would use sex selection if it were available."

There are several ways to do so, ranging from the primitive and absolutely taboo to the extremely high-tech. They include:

- infanticide
- abortion
- sperm selection
- embryo selection

Of these, the first—**infanticide**—is murder and illegal everywhere, though it has been practiced throughout history. The normal ratio of boys to girls at birth is roughly 106:100; the reason for the discrepancy is somewhat obscure, but it tends to be partially balanced by higher infant mortality among males. However, many societies have historically had imbalances that are far greater, and can only be explained by the occurrence of female infanticide. Each individual instance was likely to be called "accidental death" but the cumulative statistics tell the story. In medieval England, for example, "accidental" infant death was so predominately a female phenomenon that the child sex ratio was 130:100 (boys:girls) between 1250 and 1358.[31]

In modern times, similar results are found in some parts of the world, but the methods used to achieve them are different. Figures from the Chinese census of 2000 and the Indian one of the following year are shown in **Table 5.1**. The striking fact is that the discrepancies are greatest not in the backward, rural areas but in the *richest* parts of both countries.

TABLE 5.1
GENDER IMBALANCES IN CHINA AND INDIA[32]

REGION	BOYS PER 100 GIRLS
Natural distribution	105–107
CHINA, 2000	
National	117
Guangdong province	130
Hainan province	135
INDIA, 2000	
National	108
Rural	105
Urban	111
Delhi, citywide	116
Delhi, affluent suburb	126

For second children in China—the one-child family is an essentially urban policy—the ratio is 151 boys per girl; for third children, 159.[33] Infanticide was once widespread but is now thought to be very rare; in the 1950s, '60s, and '70s, the boy:girl ratio was both normal and stable. In modern times, however, ultrasound scans followed by **abortion** have become common even in rural areas, and clearly more so in the booming "enterprise zones"—Guangdong is the richest province in the country, and Hainan is its neighbor.

Similarly, in some of the most exclusive areas of Delhi, the capital of India, there are 126 boys for every 100 girls. Abortion for sex selection is explicitly illegal, but "wealthy housewives swap names of medical practioners willing to break the law."[34] A widely quoted 1984 UNICEF report on sex-selection abortions performed in Bombay, India's commercial center, found that of 8,000 aborted fetuses at one clinic, 7,999 were female.[35]

The American experience may be different, for a number of reasons. There is anecdotal evidence of women choosing to have girls rather than boys, which may make a difference.[36] (US families adopt somewhat more girls than boys, a ratio of about 53:47.[37]) Abortion in general is much less accepted in the US, and it is probably much harder to find a doctor who will perform an abortion for sex selection. But there are now technologies that mean you can pick your baby's gender *before* you get pregnant. And they are beginning to be openly advertised in mainstream media like the *New York Times*, which regularly runs them in the Sunday Styles section.[38]

■ SPERM SORTING ■

THE PRINCIPLE BEHIND sperm sorting is simple to the point of being obvious. When males produce sperm, each sperm includes one and only one of the two chromosomes a human has, the other being an X chromosome supplied by the female's egg. Half of the sperm have an X chromosome (so the resulting embryo would have two, and be female), half a Y chromosome, which would combine with the female's X chromosome to produce a male.

Normally, both are equally included among the millions released at each ejaculation, so the odds are pretty much even. But a relatively new technique, developed for livestock breeding (see **Box 5.7**), allows technicians to separate the X from the Y chromosome sperm, with fair success. Artifical insemination—not particularly high-tech—can then be used to inseminate the woman, and with luck to induce pregnancy.

5.7   MICROSORT®—YOUR TAX DOLLARS AT WORK

MICROSORT SPERM SORTING technology was invented in the late 1980s by Lawrence Johnson, a scientist working for the United States Department of Agriculture, as a way to pick the sex of livestock.[39] In 1992, the government patent was licensed exclusively, until it expires in 2009, to the Genetics & IVF Institute, for an unknown sum. (The technology transfer program of which this is part raises $2.6 million a year from 221 licenses.)

The process relies on the fact that the X chromosome in humans is about 2.8 percent larger than the Y chromosome. Sperm are stained with a fluorescent dye and then a laser is shone on them; the X chromosome reflects more light. The process is simpler with animals in which the difference is greater (about 4 percent in cattle), but has been refined to the point where the success rate for identifying females is 91 percent; it is 73 percent for identifying males.[40]

Fees for the service are at least $2,300, not counting travel to the clinic. The sperm is donated on the morning of the day when the woman is thought to be ovulating, and injected that afternoon. Overall, about 21 percent of customers achieve pregnancy. The average customer makes three attempts.

The first license for doing this with humans was granted specifically to prevent certain genetic diseases that are sex-linked. These mostly affect males (because females have a genetic safety net in the form of a second X chromosome) and include hemophilia and muscular dystrophy.

Rapidly, however, the clinic (the Genetics & IVF Institute in Fairfax, Virginia) began to get requests for the service from healthy couples, and the license was broadened to include any human use. The clinic now offers the service for "family balancing"—the choice of a girl as a boy's younger sibling or vice versa—but not to childless couples. (The logic behind this distinction is presumably that it does not reflect a bias in favor of either gender, though it does of course reflect a bias in favor of parental selection.)

Only about 15 percent of customers are trying to avoid diseases. Of the rest, initial inquiries are split about 50-50 between those seeking boys and those wanting girls, but 80 percent of births are female.[41] This seems to be partly because the chances of getting a boy on demand are lower, so presumably many of those couples do not in the end use the service.

One way around this would be to use a more invasive, more expensive procedure that is at present frowned upon by the American Society for Reproductive Medicine for simple sex selection—embryo testing. And this is where human genetic engineering really comes in.

■ EMBRYO TESTING ■

THERE IS A possible medical reason (unconnected with infertility) to consider using technological reproduction. In theory, and to some extent in practice, ART enables doctors to test the embryos, before implanting them in the mother, in order to ensure that the resulting children are not born with serious genetic defects.

This is an extension of what has been routine for several decades, in the form of testing during pregnancy, whether by ultrasound, amniocentesis, or other means. These tests can identify a number of serious diseases, including Tay-Sachs, cystic fibrosis, and Down syndrome, but the only "treatment" available is abortion. If you can test the embryo before the pregnancy even begins, however, there is no need for an abortion; that embryo is simply not used.

Some people do consider this selection a form of abortion, and oppose the procedure on principle. Others, notably some disability rights activists (see **Chapter 11**), are concerned about the prejudice demonstrated in choosing not to have a child whose condition may even not be truly debilitating. (While Tay-Sachs disease is incurable, ghastly, and fatal at a very young age, Down Syndrome is variable in its effects and some people live relatively normal lives.)

Certainly, such selection is a crude, small-scale, personal form of eugenics (see **Chapter 8**). Nevertheless, for those who would choose to abort a Down syndrome fetus, **Preimplantation Genetic Diagnosis** (**PGD**; see **Chapter 2** for more scientific details) seems like a better alternative—if they are undergoing IVF anyway.

That is why, as Melissa Healy wrote in the *Los Angeles Times*,[42] "For the nation's roughly 400 fertility doctors, PGD . . . brings in a new and fertile class of patient (genetic disease carriers) . . ." Genetic disease carriers are people who have one bad copy and one good copy of a particular gene, and do not suffer symptoms, because the bad gene is recessive, that is, it defers in effect to the good one. If they have a child with someone who has the same condition, then by very simple genetics there is a 25 percent chance that the child will inherit the bad copy from both parents, and suffer the disease. So, if you and your partner have both been tested and identified as carriers . . . do you want to take that chance?

That's the basic pitch. There are other possibilities as well—

Down syndrome tends to afflict the children of older women (for whom unfortunately fertility treatment is also generally much less successful); environmental exposure could be a concern, perhaps for troops returning from warfare in which depleted uranium has been used (this is controversial); any identifiable risk factor could trigger a recommendation to have your potential children screened.

Even without PGD, this approach can work. Incidence of Tay-Sachs disease fell by 95 percent over thirty years among the group that used to be most susceptible to it, Ashkenazi Jews, because a systematic program was instituted to test and warn potential carriers.[43] Some couples chose not to marry, others not to have children or to test and abort if necessary.

With PGD, fertility specialists hope to increase their target market dramatically. The same one who sees the industry at the "core of capitalism" has said,[44] "In the future, every single treatment will involve some chromosomal analysis. . . . The important present feature of PGD is its expansion to a variety of conditions which have never been considered as an indication for prenatal diagnosis, including the late-onset disorders with genetic predisposition and preimplantation non-disease testing."

■ PLAYING THE ODDS ■

THERE IS, HOWEVER, a problem for these rosy business scenarios. Genetic tests generally do not give definite answers. They tend to indicate a statistically significant difference in risk level, rather than a straight prediction. This is not always the case—see **Chapter 2** for further discussion—but it certainly complicates any such decision making.

To illustrate this, consider the widely examined "breast cancer test" developed and publicized by Myriad Genetics. This identifies mutations in the genes BRCA1 and BRCA2, which are associated with some 5–7 percent of breast cancers. If you have the mutation, you are more likely to get cancer; but at least one-fifth, maybe 40 percent of those who do have the mutation (in other words, are "at risk") do *not* develop breast cancer. And the vast majority of women with breast cancer do *not* have the mutations.[45] The test may be valuable, especially in encouraging regular checkups, but it is a long way from definitive.

Moreover, in the foreseeable future, as more tests are developed and more genetic probabilities are identified—there are said to be over six thousand conditions associated with "an error in a single gene"—balancing the odds is inevitably going to become extremely difficult.[46] It is hard to sell something that looks remarkably similar to guesswork . . .

Still, some tests *are* definitive: The extra chromosome associated with Down syndrome is relatively easy to identify. So is the Y chromosome. Which brings us right back to sex selection.

Sex selection *can* be done with PGD. It's technological overkill (and the subsequent pregnancy may not succeed). It's financially absurd (except for the seller), though that has not stopped some couples from flying across the country to one of the few clinics that sells the service.[47] But it's a foot in the door for those who want to expand the market for ART services from medical treatment—whether for infertility or to avoid diseases in the children—into the purely elective realm.

As one practitioner—who refuses to use PGD for sex selection—wryly noted, "The last time I checked, sex was not a disease."[48]

■ A GENETIC ARISTOCRACY? ■

WHY WORRY ABOUT all this? We already live in a class-ridden society where, as Katha Pollitt once put it, yuppie control freaks prepare for conception "by giving up cigarettes and alcohol and unhealthy foods, reading Stendhal to their fetuses in French."[49] What's one more manifestation of the market?

That is the argument put forward rather melodramatically by Professor Lee Silver, who envisages a future genetic aristocracy (see **Box 10.3** for more details). The attempt to portray Human GE as the inevitable result of market forces is the *reductio ad absurdum* of the market philosophy, the extreme that makes the whole principle seem ludicrous. But that is exactly what advocates such as Silver are saying; indeed, he comes very close to saying, with unassailable logic but a startlingly false premise, that the market has already impartially judged who is most fit by supplying them with the most money.

This may seem like a paranoid fantasy, but it's taken from a best-selling book, by an Ivy League professor, and promulgated in supposedly reputable intellectual circles (see **Chapter 10**). Silver and Stock and their cohorts are performing an interesting piece of intellectual juggling. They are claiming that Human GE will happen because the market will compel it, and justifying this by pointing at the virtually unregulated activities of today's ART industry.

Sometimes they claim that mass production will bring costs down, so that every home will have not only a DVD player but a GE baby (both being commodities thought to be of value). At others, perhaps noticing that inequality in the US has been growing for a generation, they lean to the genetic aristocracy line. Either way, they are assuming that technologists can overcome people's opposition to Human GE (see **Chapter 9**) with a concerted sales pitch.

■ MARKETING FAILURES ■

THE GE FOOD INDUSTRY has been lobbying hard to prevent labeling. In other words, they want to sell the product without admitting what they are doing. They know, from all the polls, that a clear majority of people prefer not to buy GE food (details are in **Chapter 9**). That's even true in the US; in Europe the numbers are overwhelming.[50] EU rules insist that GE food be labeled, and that alone is enough to keep them off the market—consumers don't buy them.[51] What is sometimes forgotten, however, is that the industry began by trying to sell its products directly to consumers. It didn't just admit that they were GE, it boasted that they were.

The FlavrSavr tomato was spectacularly ill-named. It tasted terrible—bland to the point of being a complete waste of time. Rats refused to eat it, showing great good sense. This is what happened when it was force-fed to them: "Several developed stomach lesions; seven of forty died within two weeks."[52] Nevertheless, it was approved for sale to humans on May 18, 1994. We didn't like it either. It was taken off the market in 1997.

There has been an equally notable failure in the human reproduction business—the **Nobel sperm bank**. Robert K. Graham, a southern Californian real estate millionaire, had the bright idea that if Nobel Prize winners (men, of course) would donate their sperm, or sell it, customers would line up around the block for the chance to use it. He supposedly lined up three or five (reports vary) Nobel laureates and founded the "Repository for Germinal Choice" in 1979.[53]

The plan was explicitly eugenic (see **Chapter 8**)—he insisted that the *buyers* be at least Mensa members. He had to lower his standards for "donors" when he couldn't attract enough laureates (besides, old guys' sperm just isn't as good as young men's), but he insisted they be Mensa members and fill out an application with questions such as "Have you ever had delusions of greatness or omnipotence?"[54] (It is not clear what the correct answer to that was.)

There was a curious lack of demand. It is said that 240 or so babies were eventually born from that particular sperm bank, and Graham kept it going till his death in 1997, but it closed not long thereafter.

Maybe it *is* possible to go broke underestimating the American public.

■ FURTHER READING ■

Free Documents from the Web

Our Bodies Ourselves (OBOS) is active in several related areas, including the issues of fertility drugs and direct-to-consumer advertising of drugs and medical procedures, not necessarily conducted with reproductive services; see www.ourbodiesourselves.org/lupron.htm and www.ourbodiesourselves.org/dtca3.htm. The site has good links, too.

Marcy Darnovsky, "Revisiting Sex Selection: The growing popularity of new sex selection methods revives an old debate," *Genewatch*, 01/03; www.gene-watch.org/genewatch/articles/17-1darnovsky.html

The Centers for Disease Control and Prevention (CDC) *Assisted Reproductive Technology Reports* include annual lists of ART success rates and a general overview of the industry, from www.cdc.gov/reproductivehealth/art.htm. They are available as web pages or large (over 2mb) pdf files.

The American Society for Reproductive Medicine (ASRM) website, www.asrm.org, includes a number of fact sheets for patients, as well as information for professionals.

The Society for Assisted Reproductive Technology (SART) has over 370 practice members, representing more than 95 percent of the ART clinics in the US, according to its website, www.sart.org, which also includes the 65-page document *A Patient's Guide to Assisted Reproductive Technologies*, available as a pdf (165k) or very large single web page.

Books

Lori Andrews, *The Clone Age*, Henry Holt and Company, 1999, is the most easy-to-read, and yet authoritative, book on any of the issues related to Human GE and, despite its title, largely about the ART industry.

Gregory Stock, *Redesigning Humans: Our Inevitable Genetic Future*. Houghton Mifflin, 2002. The 2003 paperback edition, published by Mariner Books, has a different subtitle: *Choosing Our Genes, Changing Our Future*.

Living with the Genie: Essays on Technology and the Quest for Human Mastery, ed. Alan Lightman, Daniel Sarewitz and Christina Desser, Island Press, 2003

■ ENDNOTES ■

1 Gregory Stock, *Redesigning Humans: Our Inevitable Genetic Future*. Houghton Mifflin: New York, 2002, p. 55

2 Amy Tsao, "Genetic Testing Meet Mad Ave," *Business Week*, 07/28/04

3 Reuters, 08/03/04

4 AP, 08/13/04

5 Marcy Darnovsky, "Revisiting Sex Selection: The Growing Popularity of New Sex Selection Methods Revives an Old Debate," *Genewatch*, 01/03

6 Lee M. Silver, "Can You Make My Kid Smarter?" (cover story), *Time*, 11/08/99, p. 68

7 "Got Eggs?" was the headline of a small ad in the *SF Weekly*, San Francisco, CA, 05/12–18/04. The ad offered $5,500 for egg "donors" and $22,000 for surrogates.

8 Martha Frase-Blunt, "Ova-Compensating? Women Who Donate Eggs To Infertile Couples Earn a Reward—But Pay a Price," *Washington Post*, 12/04/01; Jessica Cohen, "Grade A: The Market for a Yale Woman's Eggs," *The Atlantic Monthly*, 12/02

9 See www.ourbodiesourselves.org/lupron.htm

10 Debra Pickett, "Sunday Lunch with . . . Yury Verlinsky," *Chicago Sun-Times*, 08/22/04

11 Lori B. Andrews, "Changing Conceptions: Governance Challenges in the Engineering of Human Life" in *Living with the Genie: Essays on Technology and the Quest for Human Mastery*, ed. Alan Lightman, Daniel Sarewitz & Christina Desser, Island Press, 2003

12 www.cdc.gov/reproductivehealth/art.htm

13 The President's Council on Bioethics, *Reproduction and Responsibility: The Regulation of New Biotechnologies*, Washington, D.C., 2004, p. 155

14 P. Katz, R. Nachtigall, and J. Showstack, "The Economic Impact of the Assisted Reproductive Technologies," *Nature Cell Biology & Nature Medicine*, Fertility Supplement, 2002, s29–32, citing a study by J. Collins, *Semin. Reprod. Med.* 19, 279–289 (2001).

15 The President's Council on Bioethics, op. cit., p. 43

16 One study indicated a greatly increased risk of Beckwith-Wiedemann syndrome (BWS) but the million babies studied included only four that were conceived with IVF and suffered from BWS. BBC, 08/14/04.

17 Transfer rates vary, on average, from about 2.8 to 3.7, depending on several factors including the woman's age and whether the eggs used were donated; US Department of Health and Human Services, *2001 Assisted Reproductive Technology Success Rates*, p. 71

18 Dr. Cecil Jacobson, quoted in Lori Andrews, *The Clone Age: Adventures in the New World of Reproductive Technology*, Henry Holt and Company, New York, 1999, p. 78

19 AP/*Los Angeles Times*, 08/04/04

20 Andrews, op. cit., pp. 76–80

21 Andrews, op. cit., p. 53

22 ibid.

23 personal communication from Luke Anderson, who was there (at the Englesberg Seminar in Sweden, organized by the Axxon Johnson Foundation) and took notes.

24 www.extendfertility.com

25 BBC, 01/16/05; *Philippine Daily Inquirer*, 04/16/97

26 *2001 Assisted Reproductive Technology Success Rates*, op. cit., p. 71

27 Kristen Philipkoski, "Frozen Eggs Showing Promise," *Wired*, 09/13/04

28 Alex Mar, "Freezer Burn: Is a Procedure That Allows a Woman to Preserve Her Eggs All It's Cracked Up to Be?" *Slate*, 08/12/04

29 Our Bodies Ourselves continues to monitor this issue; see in particular www.ourbodiesourselves.org/lupron.htm.

30 Meredith Waldman, "So You Want a Girl?" *Fortune*, 02/19/01, Vol. 143 Issue 4, p. 174

31 Marvin Harris, *Cannibals and Kings*, Vintage Books: New York, 1978, p. 258, citing Josiah Russell, *British Medieval Population*, Albuquerque: University of New Mexico Press, 1948.

32 Chinese data from John Gittings, "Growing Sex Imbalance Shocks China," *The Guardian* (London), 5/13/02. Indian data from *Provisional Population Totals: India*, Census of India 2001, www.censusindia.net.

33 Robert Marquand, "China Faces Future as Land of Boys," *Christian Science Monitor*, 09/03/04

34 Catherine Philp, "Delhi's Rich Adopt Gender Selection of the Poor," *The Times* (London), 11/27/02

35 Cited at www.gendercide.org/case_infanticide.html and elsewhere; I have not been able to locate the original report. It is often misquoted to imply that only females were ever aborted; what it says is rather that only females were aborted because of their gender, which is not quite the same thing.

36 An Australian clinic reported that 55–60 percent of couples requested girls. Australian ABC, 08/30/04

37 *Forbes*, 04/27/04

38 Marcy Darnovsky, "Sex Selection Moves to Consumer Culture," *Genetic Crossroads*, 8/20/03

39 Waldman, op. cit.

40 "Seattle Gathering of Fertility Experts Spotlights Gender-Picking," *The Seattle Times*, 10/16/02

41 *Seattle Times*, op. cit.

42 Melissa Healy, "Fertility's New Frontier," *Los Angeles Times*, 7/21/2003

43 Gina Kolata, "Using Genetic Tests, Ashkenazi Jews Vanquish a Disease," *New York Times*, 2/18/03

44 Christine Rosen, "Eugenics—Sacred and Profane," *The New Atlantis*, Summer 2003

45 Barbara Katz Rothman, "Cancer Is (Not) a Genetic Disease" in Brian Tokar, ed., *Redesigning Life? The Worldwide Challenge to Genetic Engineering* (London & New York: Zed Books, 2001); Denise Grady, "Tests for Breast Cancer Gene Raise Hard Choice," *New York Times*, 03/05/02

46 Julie Robotham and Deborah Smith, "Looks Who's Perfect Now," *Sydney Morning Herald*, 08/28/04

47 Beth Whitehouse, "PGD: Doctors Reluctant," *Newsday*, 06/14/04

48 Julie Bell, "Technology Customizes Kids By Sex," *Baltimore Sun*, 05/10/04

49 Katha Pollitt, "It's a Bird, It's a Plane, It's . . . Superclone?" *The Nation*, 07/23/01

50 ABC News poll, 07/03; *Guardian*, 05/07/02; "Consumers clearly don't want GM food and are hardening their stance against it," Malcolm Coles, editor of *Which?* magazine, commenting on a UK poll, 09/04, reported at www.foodnavigator.com/news/news-NG.asp?id=54550

51 Elisabeth Rosenthal, "EU Labeling Laws Have Driven GE Foods off the Market," *International Herald Tribune*, 10/05/04

52 Jeffrey M. Smith, *Seeds of Deception*, Yes! Books, Fairfield, IA, 2003, p. 157; see also pp. 137–140, 239.

53 David Plotz, "The 'Genius Babies,' and How They Grew," *Slate*, 02/08/01, and several subsequent issues up to 03/30/01.

54 Lori Andrews, *The Clone Age*, op. cit., pp. 124–139. She posed as a potential client, taking the Mensa test specially, so now we know for sure she is smart, or at least has a high IQ, or at the very least can pass the Mensa test.

6

GENE THERAPY— "BETTER THAN WELL"?

■

INTRODUCTION

GENE THERAPY IN THE 1990s:
FROM PANACEA TO DISASTER

THE NEXT FALSE DAWN

OF MARKETS AND MEDIA

LEARNING THE WRONG LESSONS

IN UTERO SOMATIC GENE TRANSFER PROPOSAL

THE INDIVIDUAL AND SOCIETY

FURTHER READING

■

■ INTRODUCTION ■

I F A BAD GENE were to cause a disease, and you could replace the gene, wouldn't that be great? You'd have not just a treatment but a *cure*. And what if you had inherited an adequate but less-than-ideal gene? Swap that one out and you'd be . . . better than well![1]

Things are not nearly so simple (see **Chapter 2**) but that's the principle behind the hope of what is called "gene therapy"—and the perhaps more imminent hazard of "gene doping" in sports (see **Chapter 7**). Some people even seem to think these techniques may lead to eternal, or at least extended, life.

"Gene therapy" is really a misnomer: So far, there have been hundreds of experiments but very little success and no *therapies*. The term is also used rather loosely to cover several different kinds of interventions, some of which involve not *replacing* genes but turning them on or off.[2] Still, there is enormous power in a simple and comprehensible story (see **Chapter 4**, on stem cells), and the term is entrenched. Even if it is somewhat misleading, there's no avoiding it.

The very failure of gene therapy ironically *increased* the pressure from some quarters to attempt the full-fledged human genetic engineering of children, as discussed in more detail below. If *somatic* interventions (see **Box 6.1**) are hard to control because you have to change so many cells, then *germline* ones, in which you only have to modify one single cell (or very few) might be more achievable.

6.1 SOMATIC VS. GERMLINE GE

THERE IS A crucial distinction between somatic and germline genetic engineering, or "therapies" as advocates of Human GE like to call them. It is discussed in more detail in **Chapter 2**, but briefly:

> **Somatic** genetic engineering affects some of the cells in a single body but is not passed on to future generations because it does not change the eggs or sperm.

> **Germline** genetic engineering does affect the eggs or sperm and is therefore passed on to future generations, who will carry the alterations in every cell of their bodies.

"Germ cells" are eggs, sperm, and the cells that make them; they pass on genes to offspring, and this shared lineage from parent to child is known as the "germline" of the organism or, more broadly, the species—the continuous inheritance from ancestor to descendant.

"Somatic" comes from the Greek for "body" and may be used to contrast either with the psyche or with the germ cells, depending on context. In GE contexts, somatic interventions only affect one particular body.

Intentional Human GE would of course be done in the lab, using very early embryos and IVF techniques, not the germ cells of adults. However, because gene transfer is not an exact science, some kinds of intervention that are meant to be somatic may also, by mistake, affect the germ cells. If they do, they could be passed on to children and thus affect the germline.

One controversial proposal that admittedly risks this, perhaps in a deliberate attempt to push the envelope, is discussed later in this chapter.

There are those, led perhaps by James Watson (see **Box 10.1**), who think this is a good idea. Some of them are explicitly in favor of eugenics, while others, including Watson, fail to comprehend the eugenic implications of their well-meaning ideas (see **Chapter 8**). There are also those, however, like Lee

Silver and Gregory Stock, who appear not to care (see **Chapter 10**). Typically, they take a libertarian view that essentially denies the right of society as a whole to influence or restrict their personal decision making. This, in turn, may include a decision in favor of "enhancement" (if that is indeed possible)—gene doping for sports, albeit at unknown long-term risks, or attempts to increase lifespan, attractiveness, memory, and so on.

Most of the research thought to be leading to these possibilities is justified on medical grounds. There are, however, legitimate reasons to question these priorities. Some of them are scientific: The predictability of genetic interventions remains quite low, and given the complexity of the system they are trying to influence may never work reliably (see **Chapter 2**). Others involve the ethical questions raised by experimentation on humans (see **Chapter 3**). And others again are social, and raise questions about the priorities we collectively choose to stress and, especially, fund.

Just as each member of a germline is a unique individual, so is each a member of society. Improving the health of one, as long as it doesn't hurt anyone else, improves the health of society as a whole. But some kinds of social interventions can benefit society as a whole without being aimed at any given individual—and some uses of resources that benefit a few may indirectly harm a much greater number of people. This chapter ends with a brief discussion of this critical dilemma.

■ GENE THERAPY IN THE 1990s: FROM PANACEA TO DISASTER ■

IN 1990, gene therapy was the Next Big Thing. W. French Anderson and his colleagues at the National Institutes of Health (NIH) began the first clinical trial on September 14, 1990. In the next few years, they and others started several hundred more. This, they thought, was the medicine of the future.

In a few years, however, even Anderson, who is routinely called "the father of gene therapy," had to admit scientists "came to realize that nothing was really working at the clinical level."[3] A review in *FDA Consumer*, the Food and Drug Administration magazine, summed up what happened in 1999. The field was approaching a dead end, "And then things got worse."

A patient died, Jesse Gelsinger (see **Box 6.2**).

6.2 **THE JESSE GELSINGER TRAGEDY**

JESSE GELSINGER WAS an eighteen-year-old volunteer for a gene therapy trial at the University of Pennsylvania, who had a severe adverse reaction and died four days later, on September 17, 1999. His death was a tragedy. When the details began to come out, however, it began to look more like a scandal.

Gelsinger suffered from a rare liver disease that meant his body could not remove ammonia. He had a mild version, which could be controlled with diet and drugs, but he did have to take thirty-two pills a day.[4] The therapy aimed to introduce corrective genes that would generate the missing enzyme for metabolizing ammonia. This treatment would not necessarily cure his condition, but might help, and if it worked, could potentially be lifesaving for newborns with a more severe version. He was quite aware of this, which made him in some ways an ideal candidate; he had told a friend,[5] "What's the worst that can happen to me? I die, and it's for the babies."

Physically, however, he should have been disqualified; his blood ammonia levels were too high.[6] Also, he could not give truly informed consent because he was not told that:

- Four previous patients had become sick
- Two monkeys had died from a similar gene drug

Neither event had been reported to the FDA, as they should have been, and either one by itself should have led to the trial being halted.[7]

The test went ahead. On Monday, September 13, Gelsinger was injected with 30 milliliters (two tablespoons) of the material that carried

6.2

the genes that should theoretically have cured him. That night he ran a fever of 104.5°F. The next morning, he was jaundiced. That afternoon he was in a coma. That night he was on dialysis. He suffered multiple organ failure, his lungs stopped working and on Friday they pulled the plug.

At that point, Paul Gelsinger, Jesse's father, was sympathetic to the scientists. He changed his mind, "concluding that he had been duped by scientists who cared more about profits than safety" and filed suit a year later.[8] The legal action was quickly settled, for an undisclosed sum, estimated at $10 million.[9] But, he said,[10] "Dealing with the money end of this [settlement] was probably one of the most difficult aspects of this because this experiment was all about money, and it was never about money for Jesse."

The conflicts of interest—not just potential but actual—were startling. Dr. James Wilson, the lead scientist and director of the Institute for Human Gene Therapy at the University of Pennsylvania, had founded a company called Genovo, which stood to profit if the experiment worked.[11] Both he and the university owned stock; the man who had hired Wilson, William Kelley, the dean of the medical school, was another major stockholder.

Wilson was banned by the FDA from participating in human research, and resigned as director of the institute.[12] He remains a professor at the university, however, and his experiments on monkeys have raised fears about the possibilities of gene doping by athletes (see **Chapter 7**).

And then it came out that at least ten other experimental subjects had died—and some, if not all, of their deaths seemed to have been covered up, or at least inadequately disclosed. These all occurred *before* Gelsinger's death, though his was the first publicly known:

- **Four patients died** during gene-therapy experiments headed by Dr. Ronald Crystal of the New York Hospital-Cornell Medical Center in Manhattan.[13]
- **Two patients died** in the course of gene therapy performed at St. Elizabeth's Medical Center in Boston, by Dr. Jeffrey Isner of Tufts University.[14]

▶ **Three of the first six patients died** in a gene-therapy trial conducted by Harvard Medical School researcher Dr. Richard Junghans at Beth Israel Deaconess Medical Center, also in Boston; the next one got violently ill but survived.[15]

▶ **Another died** in Toronto in 1997, as a result of a gene-therapy experiment for brain cancer.[16] An Indiana patient who was part of the same, Novartis-sponsored experiment is also reported to have died just before the Canadian, James Dent, entered the second, fatal, phase of his treatment.[17]

Crystal, Isner, and Junghans all concluded that the deaths of their patients were probably not related to the gene treatment—the subjects were very ill already—and none reported them to the National Institutes of Health (NIH) for public review, as required. (Dent's death should have been reported to Health Canada within a week; the autopsy report was eventually sent after five months.[18]) At least one of them, Junghans, did report the event to the FDA, and at least one, Crystal, did tell the NIH but asked them to keep the information confidential until he investigated further. This was just after his biotech company had filed for an initial public offering of stock, though Crystal later denied any connection. The stock offering never happened, and by the time the news came out in the regular press he had already published a report in an academic journal and discussed the deaths at scientific meetings.

After an emergency request for filings, the NIH discovered that less than 4 percent of "serious adverse events" (37 of 970) occurring in gene transfer trials were being reported as required.[19] "Serious adverse events" means exactly what it says—death or sickness, in patients undergoing gene therapy, whether or not the illness appeared to be connected with the trials. As Amy Patterson, head of the NIH Office of Recombinant DNA Activities (ORDA), explained:[20] "It may take five, six, seven patients ill, or

20 patients, before you find out, 'Hey, this is also happening in other people's trials.' And if you don't know what's going on in other people's trials, then you can't put two and two together."

When the data did get reported, the details were deeply troubling. There were many instances of fevers and drops in blood pressure, similar to Gelsinger's experience. There were also repeated examples of problems that seemed to have been caused by attempts to deliver genes into patients' brains— difficulties with speech and occasional paralysis.

An even more disturbing pattern emerged when the deaths of patients involved in separate gene therapy experiments were considered together. This is how the *Washington Post*'s Deborah Nelson and Rick Weiss, who did excellent work chasing and often breaking this story, characterized what they found:[21] "Throughout the reports to the NIH, scientists repeatedly acknowledge that their gene treatments caused various non-lethal symptoms, but they invariably conclude that any deaths likely resulted from underlying illnesses or other causes"— which may have been true, in any individual case. The patients were very ill, probably terminal, so the doctors had what one researcher Nelson and Weiss spoke with called "plausible deniability." When all the reports are taken together, however, the evidence—while circumstantial—becomes hard to ignore, as when, in Thoreau's happy phrase, you find a trout in the milk.[22]

■ THE NEXT FALSE DAWN ■

NEVERTHELESS, THE CONCEPT of gene therapy was so fascinating that work on it continued, despite all the disasters. New rules were put in place, old ones were enforced, and rather soon—in France, as it happened, but just in time to revive the field—there was a breakthrough.

A few babies—perhaps one in a million—are born with essentially no immune system, a condition known as Severe

Combined Immunodeficiency (SCID).[23] This is a cluster of different disorders with similar consequences: The patient cannot fight infections and, if untreated, generally dies in the first year, unless kept in a completely sterile environment, a "bubble." Most of them are boys, since a girl with a harmful mutation on an X chromosome can compensate if the other is normal (see **Chapter 2**); hence the subgroup, X-SCID.

Bone marrow transplants can help, because the stem cells in bone marrow ("adult" stem cells, not embryonic; see **Chapter 4**) generate the blood which contains the various immune system cells. This does require a suitable donor, and there is not always one available.

Scientists working at the Necker Children's Hospital in Paris found another solution for boys with a specific bad gene: They took bone marrow from the babies (for the first two, when they were eight and eleven months old), isolated the stem cells, and then genetically engineered them.[24] They did this by infecting them with a kind of virus to which they had attached a working copy of the relevant gene. They hoped this would replace the bad one in these stem cells. Then they put the altered stem cells back. And it worked! The altered stem cells began to generate blood that did contain the replacement gene—within fifteen days, scientists could detect it in the blood.

Three months later, the boys left their bubbles and went home. After ten months, several types of immune system cells had reached normal levels, and both children were doing fine. The report was published in *Science* in April 2000.

Nine more babies were treated at Necker, almost all successfully, and scientists at London's Great Ormond Street treated seven more.[25] This was *the* big success of gene therapy, at long last.

Unfortunately, two of the boys got sick, though not immediately. One showed symptoms of leukemia (vast overproduction of white blood cells) about two and a half years after the treatment, in October 2002. A second developed symptoms the following January.

What seems to have happened is that the virus that brought in the good gene left it in a bad location within the patient's DNA. It was inserted next to, and apparently turned on, a gene that eventually triggers leukemia ("insertional mutagenesis"). The first occurrence seemed like really bad luck; the second suggested that there might be a specific reason for this.[26] Alain Fischer, head of the Necker team, suggests that two relevant factors may be that these were the youngest patients, one and three months old (the others were all at least six months old) and that they were injected with the highest number of engineered cells.[27]

The FDA was promptly notified and, in January 2003, suspended the twenty-seven most similar experiments in the US, which involved several hundred patients.[28] (These trials all used retroviruses to introduce replacement genes to blood stem cells; about two hundred other experiments, using somewhat to very different methodologies, were unaffected by the moratorium.[29])

One of the baby boys with leukemia died in October 2004, but the other responded well to chemotherapy and bone marrow transplants. Meanwhile, British scientists had successfully treated seven more, and the French and US trials were cleared to resume. Fisher explained,[30] "No therapy is without risk, and . . . this therapy, even with the risk, may be better than the current treatment." But in January 2005, a third child developed leukemia.[31] They don't know why. The trials were halted again.

For the families of those few who were successfully treated, gene therapy seems like a miracle cure. Turning that into routine medicine, however, is still very far away.

■ OF MARKETS AND MEDIA ■

THE GRANDIOSE FANTASIES of gene therapy have given way to something much closer to basic science, but that's not always the way the media portrays them. There is still an allure to the concept, and journalists like to apply it to such important—or at least potentially profitable—applications as baldness cures (see **Box 6.3**), even when the term is not exactly relevant.

6.3 HI-TECH TREATMENTS FOR BALDNESS

BALDNESS MUST HAVE been one of the first "nonillnesses" to get treated: The ancient Romans had a formula that included wine, saffron, pepper, vinegar, a medicinal herb and—no doubt crucially—rat dung.[32] (Presumably the treatment did not work.)

Hair loss is still big business—by some estimates worth $1 billion a year in the US alone. In the US, 40 million men and 20 million women have lost at least some of their hair.[33] If even half of them want it back . . . well, what entrepreneurial scientist wouldn't like a piece of that market?

That may sound cynical, but the 2003 publication, in *Nature*, of a significant study by Dr. Elaine Fuchs about the adult stem cells in the hair follicles of mice was accompanied by a press release from the Howard Hughes Medical Institute with this pull quote:[34] "These studies raise the possibility that drugs to activate these natural factors could promote hair follicle growth." A *New Scientist* headline in March 2004 was "Stem cell tricks hint at baldness cures."[35] By September 2004, *Nature* was reporting that "Stem cells could allow people to grow a new head of hair."[36] No wonder *Forbes* carried the story.[37]

There is important basic science here. What it is not, however, is gene therapy. In fact, the breakthrough reported in September 2004 was precisely the opposite, that "this new methodology for isolating skin stem cells does not require genetic manipulation," which therefore "opens the door for applying this method for isolating human cells."[38]

Of course, another "treatment"—guaranteed 100 percent reliable, affordable to all, and with no side effects—is not to worry about it.

This would not be the first time that complex and even poisonous technologies have been applied to the disease of vanity. Plastic surgery is now so common that according to the American Society of Plastic Surgeons almost 9 *million* cosmetic procedures were performed in 2003.[39] That does not include reconstructive procedures, for burn victims and others. It does, however, include almost 3 million people who had injections of what Bill McKibben, with some disbelief, points out are dilute strains of the botulism toxin that his mother warned him about getting from dented tin cans.[40] At least they are (almost all) adults. Some children also get expensive (and possible dangerous) medication for non-illnesses, such as being short (see **Box 6.4**).

6.4 SHORT PEOPLE

HUMAN GROWTH HORMONE is a substance that some athletes abuse. So do thousands of innocent children. It was developed to treat the estimated 2,500 people in the US who suffer from pituitary dwarfism, a condition in which the body fails to make its own hormone.[41] Therefore, the manufacturers were given "orphan drug status," a guaranteed seven-year monopoly intended to encourage the development of low-market drugs. Genentech and Eli Lilly both got this, for Protopin and Humatrope, respectively.

Then they began marketing it to short people. Once a drug is approved, any doctor may prescribe it. By 2003, sixteen years after its introduction, 200,000 children worldwide had taken Humatrope. Then the FDA agreed to allow its use for "children who are healthy but abnormally short"—boys predicted to be less than five-foot-three, girls predicted to be less than four-foot-eleven; the shortest 1.2 percent.[42] The company "pledged tight restrictions" on distribution, and said that only about 10 percent of the 400,000 eligible children were expected to get treated.

Why? Presumably because short people have a hard time, but the conclusion of an extensive study comparing "short normal" children with their average peers gave "only limited support for the hypothesis that short children are disadvantaged" and noted that,[43] "The short children did not differ significantly from the control children on measures of self-esteem, self-perception, parents' perception, or behavior. . . . Social

6.4 class seems to have more influence than height on children's psychological development."

As Andrew Kimbrell notes, this is a cure in search of a disease.[44] And if, as in Garrison Keillor's *Lake Wobegon*, all our children become above average, at least in height, where does it ever stop? There will always be a "bottom 10 percent" . . .

That particular "treatment" with human growth hormone, which has been abused by athletes (see **Chapter 7** for a discussion of "gene doping"), is not gene therapy in any sense, although the artificial hormone was the product of genetic technologies. Another line of research is how to affect genes not by replacing them but by switching them on or off.[45] Again, advances are being made and will undoubtedly continue to be, although the results are not yet predictable. It is surprising, however, to see media reports that seem to focus on the profit prospects of business rather than the healthcare implications for patients (see **Box 6.5**).

6.5 MEDIA APPROACHES: THE BUSINESS OF HEALTH OR THE HEALTH OF BUSINESS?

THE NEW YORK TIMES, early on September 14, 2004, featured on the front page of its website this headline from the Business section, "Method to Turn Off Bad Genes Is Set for Tests on Human Eyes."[46]

At the same time, the Health section featured, "Doctor Puts the Drug Industry Under a Microscope."[47]

The "turn off bad genes" story included quotes from scientists with business interests either in that technology or a rival process, as well as disinterested researchers. Its URL listed it as a Business story, and it was also linked from the Health page (the *Times* regularly cross-links stories) but not from the Science page.

The "drug industry" story, in contrast, was linked from Science but not from Business, although the interview with Dr. Marcia Angell, former editor in chief of the *New England Journal of Medicine*, was entirely about her

6.5 book The Truth About the Drug Companies: How They Deceive Us and What to Do About It. It is hard to avoid the conclusion that when a business success has health side effects, it's relegated to the Health section, but when a health story provides a business opportunity, it's promoted to Business.

Any survey of the science is certain to be outdated within months. That does *not* mean that cures will be available; it may mean that treatments seem *further* away. That's how science often works. No such caveats, unfortunately, inhibit a few wildly hyperbolic academics—and journalists who are witty or gullible, or perhaps both—from bandying about claims like this one:[48] "Scientific leaps could eventually let people live forever [and] gene and stem cell therapy, as well as vaccinations to reprogram the body's immune system, [will] halt aging within 25 years." Even that enthusiast, Dr. Aubrey de Grey of Cambridge University, admitted that "in a lot of these meetings [about preventing aging] there are a number of products being sold that have no merit whatsoever."

Readers need to be extremely careful. Investigations certainly continue, on many fronts, but news reports about them—especially the headlines—are often exaggerated or downright misleading. For example, the rather cumbersome title, "First Parkinson's Gene Therapy Patient Well After 1 Year Suggesting that the Therapy Is Safe And Effective" was contradicted in the third paragraph of the same article by an expert who asserted that,[49] "Because the patient is on low doses of his medication, it is not possible to determine whether the gene therapy is any better than a more aggressive drug regime . . ."

Similarly, attention grabbers such as "Alzheimer's Gene Therapy Drug Developed" should be taken with a large pinch of salt.[50] That particular story described an experiment performed at the University Medical Center in Chicago to test a product owned by Ceregene, a San Diego biotech firm. Ceregene, in a press release, called it "the first to test the company's growth factor gene therapy delivery system in a non patient-specific, 'off-the-shelf' product formulation for patients with Alzheimer's disease." Specifically,[51] "The goal of this study is to determine the safety and efficacy of this new gene therapy system. Efficacy will be measured by memory and cognitive tests, as well as brain imaging studies." In other words, they are trying. "If successful, this could be a major step toward modifying the course of the disease."[52] Well, yes. And if not, it actually also could be, if only by eliminating one line of inquiry. It's too soon to tell.

Another Alzheimer's study, likely to be significantly less profitable, suggests that you can help stave off the onset of the disease by . . . going for a walk.[53] Yet another suggests that eating fish, olive oil, fruit, and vegetables will help. A groundbreaking British study recommends drinking tea, which the lead researcher, Dr. Ed Okello, described as "particularly exciting as tea is already a very popular drink, it is inexpensive, and there do not seem to be any adverse side effects when it is consumed."[54] He then spoiled the effect by adding, "Still, we expect it will be several years until we are able to produce anything marketable."

■ LEARNING THE WRONG LESSONS ■

THE ORIGINALLY OVERINFLATED hopes for gene therapy were partly, it seems, based on an oversimplified view of the way genes work—the "gene myth" that Professor Ruth Hubbard and others have been trying to expose for many years.[55] As a position paper from the Council for Responsible Genetics explains:[56]

[There is a misapprehension] that once a gene implicated in a particular condition is identified, it might be appropriate and relatively easy to replace, change, supplement or otherwise modify that gene. However, biological characteristics or traits usually depend on interactions among many genes, and more importantly, the activity of genes is affected by various processes that occur both inside the organism and in its surroundings. This means that scientists cannot predict the full effect that any gene modification will have on the traits of people or other organisms.

The whole premise of gene therapy may have been, if not quite false, then at least badly oversimplified. One response to the failures is, therefore, to expand the various possible directions of research—to make it both more narrowly focused and more general. This is happening, and clearly desirable. Another response, however, is much more controversial.

The overoptimistic predictions *also* reflected a serious underestimation of the difficulty of inserting a specific gene in a particular place—or just a harmless place—in an individual cell's DNA, let alone that of millions of cells. Ignoring, at least for now, the basic question as to whether gene replacement is going to be useful, some researchers have been working on the practicalities of delivery systems.

One rather obvious step is to ensure delivery to the smallest possible number of cells. That's the attraction—in this context—of adult stem cells, notably the ones in bone marrow that were used in the X-SCID experiments. You don't have to engineer every cell in the blood, you engineer the cells that *make* the blood.

But that only works for (in that example) the blood. What if you want to introduce a gene into all the cells of a body? Wouldn't it be easier to do that when very few cells are involved? Which means, of course, the germ cells, sperm or eggs, or else the very early embryo. If you can affect the critical single cell of a fertilized egg, then the change will certainly be reflected in every cell of the adult that develops from that egg.

That is unquestionably Human Germline Genetic Engineering. As such, it would not be allowed in the US, under federal guidelines (see **Chapter 12**), or in any other country that has addressed the issue. But some scientists—either oblivious, or intent on pushing the limits of acceptability—are investigating the mechanisms with which to do it, and presenting their work as the investigation of a purely technical issue.

For example, there was a report in 2001 that scientists had created "designer sperm."[57] This involved an effort to modify mouse sperm, which therefore should ensure that any changes are indeed inherited by any offspring. What was added to the mouse sperm was a "reporter" gene, which was designed to be easily identified and otherwise useless. (The process is rather like adding a dye to liquid, so you can see if the colored fluid appears as expected.) As is often the case, the research, while valuable, was less successful than the headline would indicate; less than 5 percent of the offspring inherited the "reporter" gene. But Professor Ralph Brinster, the lead researcher, made the goal quite clear, "These cells are thought by some to provide the best approach should human germline gene therapy be undertaken in the future."

Another report, in 2003, drew the headline "GM changes 'bred into mice'" from the BBC, and described a method of altering embryonic stem cells and then injected them into the very early embryo.[58] Once again, "not only were the changes taken up in these animals, but also in many of the first generation of their offspring." (Note that careful "many of" qualification.) And once again human uses were discussed, although cautiously, "Despite the effectiveness of the therapy in these animals, treatments for humans are still on the distant horizon, and the new method may not prove suitable." The "effectiveness" may be questionable, but the intent is clear.

Various other techniques have been proposed for what is optimistically called "germline gene therapy." They range from artificial chromosomes to embryo cloning (see **Chapter 2**), but they all share the working assumption that not only will "gene therapy" work—which is certainly questionable—but that the way to make it work is to glide seamlessly from the ethically acceptable "somatic therapy" to the ethically far more complex "germline intervention" which constitutes undeniable Human GE.

Perhaps this is not so much learning the wrong lesson as looking for a plausible excuse. Another proposal was even more blatant.

■ IN UTERO SOMATIC GENE TRANSFER PROPOSAL ■

W. FRENCH ANDERSON, the first person to try somatic gene therapy, floated a concept that might have made him the first person to affect the germline, albeit with a certain plausible deniability. He put a "pre-proposal" before the NIH's Recombinant DNA Advisory Committee (RAC; see **Chapter 12**)— an effort to test the waters—for an *in utero* somatic gene transfer experiment. That is, he would try to "correct" the genes of a fetus that had been diagnosed with a fatal disease, and to do so before birth.

This method carried a significant risk of affecting the fetus's germ cells and thus, eventually, the human germline. Anderson not only admitted this, he asked in his cover letter to the RAC,[59] "Should the possibility of inadvertent gene transfer to the germline be considered a benefit or a risk?"

This was in 1998, when the failure of somatic gene therapy had become obvious to those in the field, but before the Gelsinger tragedy had led to the lid being blown off the stories of abuse, illness, and death. Anderson's critics were certain that he was deliberately trying to "push the envelope" on inheritable genetic modification, by not only using therapy as the justification but putting the RAC in a classic double bind, as described in an analysis by the Center for Genetics and Society:[60]

If the RAC ruled that Anderson's proposal was acceptable, it would put the US government on record as saying that germline modification was not so objectionable an event that it should stand in the way of at least some other beneficial interventions. If the RAC ruled that the proposal was unacceptable, [that] would also work towards eventual approval of germline modification, [if it gave] official US government sanction to the notion that its eventual acceptability rests on questions of patient safety rather than on overarching ethical, moral or social values.

The RAC issued a statement on March 11, 1999, unanimously calling the idea "premature" but suggesting that further studies and "a more thorough understanding of the ontogeny of human organ systems" might lead to a time when the Committee would consider such a proposal.[61]

Anderson had suggested that he intended to put forward a full proposal in 2001. That would have been an enormous development, which would certainly have drawn worldwide attention and become the focus of a global debate. But he was too late. By then, the failures of gene therapy had become well-known and the political atmosphere was not encouraging. He did not do so, and that particular protocol as an approach to germline engineering seems to be off the table.

■ THE INDIVIDUAL AND SOCIETY ■

BY ANY MEASURE, the US spends much more on healthcare than any other nation. Only eleven countries spend even *half* as much, per person; not one, not even Switzerland, which comes second, spends more than three-quarters what the US does.[62] Moreover, in most other countries, healthcare spending is largely by the state, whereas most US consumers have a more direct relationship—they pay for it, or their employer does. This clearly does not maximize utility: In at least twenty-five countries, people live longer than Americans do.[63]

The figures are even worse for infant mortality. Every major developed country does better than the US. The United

Nations reports that the US is thirty-first in the international tables.[64] And that ranking is falling—the US was about fifteenth in 1970—even though the infant mortality rate has actually been getting much better (from 20 per 1000 in 1970, to 11 in 1986, to 7 in 2002). Everyone else has been improving faster. The *San Francisco Chronicle* put it bluntly in October 2004, "Babies born in the United States are twice as likely to die as those of Sweden, Iceland, Japan, Spain or even the Czech Republic."

In poor areas, it's much worse. Even within San Francisco itself, one of the healthier cities in the country, the rate varied tremendously. In the poorest neighborhood, infant mortality was more than twice as high as in the city as a whole—at a level comparable with Bulgaria and Jamaica, which rank fifty-sixth and seventy-ninth, respectively, on the United Nations Human Development Index.

What does this have to do with gene therapy? Nothing directly—and that's the point. It's a question of healthcare and research priorities. Clearly, this is a much larger topic than can be covered in any brief overview, but it would be irresponsible to discuss high-tech, cutting-edge, futuristic visions of individualized healthcare without at least mentioning the social context.

In 2004, the Swiss company Roche Diagnostics announced what is believed to be the first commercial diagnostic test for personalized medicine. It will cost about 450 Euros (roughly $550 in 2004), which "will mean it is only an option for wealthier patients."[65] This and related topics are likely to become widely discussed in the near future.

The British Royal Society (an independent national academy founded in 1660) is conducting an investigation into the whole issue of personalized medicines. This will consider not just the feasibility of designing medication to fit an individual's genetic profile but "what it might cost, how soon it could be achieved and if it would impact negatively on the modern healthcare system."[66]

Bluntly, can we afford it? Should some people be allowed to spend whatever they choose on treatments if that reduces the availability of other treatments to most people? Should research be driven by what companies see as profitable avenues, or should the application of resources be directed by (for instance) funding decisions of the NIH? Or some combination of the two? Whatever the decision, how should it be made?

Gene therapy—scientifically and socially—still has far more questions than answers.

■ FURTHER READING ■

Free Documents from the Web

Stephen Leahy, "The Genetics Revolution Has Failed to Deliver," *Maclean's*, 09/30/02, is a succinct, skeptical overview, at www.macleans.ca/top stories/science/article.jsp?content=72658 and also re-posted at www.ngin. tripod.com and gmwatch.org, both of which are worth browsing.

The Food and Drug Administration (FDA)'s Center for Biologics Evaluation and Research (CBER) posts the text of warning letters sent to researchers, such as the one connected with the Gelsinger case, at www.fda.gov/ cber/gene.htm. Other valuable articles are at www.fda.gov/cber/info sheets/genezn.htm and www.fda.gov/fdac/features/2000/500_gene.html.

The Recombinant DNA Advisory Committee (RAC) also has some regulatory authority over gene transfers; see www4.od.nih.gov/oba/Rdna.htm.

The Council for Responsible Genetics (CRG) call for a moratorium on gene therapy trials is at www.gene-watch.org/programs/cloning/FDA-gene therapy-comments.html.

The National Cancer Institute has an overview of gene therapy, almost all of which applies to more than just cancer, at http://cis.nci.nih.gov/fact/ 7_18.htm.

The Genetics and Public Policy Center also has an introduction, at www.dnapolicy.org/genetics/transfer.jhtml.

The American Society of Gene Therapy (ASGT) remains hopeful. Its website, www.asgt.org, carries position statements and even offers golf shirts, in medium or large, for only $25 plus $5 shipping.

▪ ENDNOTES ▪

1 This irresistible phrase was reportedly coined by Peter Kramer; it was adopted by Carl Elliott as the title of a 2003 book, which covers a much broader spectrum of "enhancement technologies," including drugs, cosmetic surgery, and psychological issues.

2 Andrew Pollack, "Method to Turn Off Bad Genes Is Set for Tests on Human Eyes," *New York Times*, 09/14/04

3 Larry Thompson, "Human Gene Therapy: Harsh Lessons, High Hopes," *FDA Consumer*, 10–11/00

4 Sophia Kolehmainen, "The Dangerous Promise of Gene Therapy," *Genewatch*, 02/00

5 Sheryl Gay Stolberg, "The Biotech Death of Jesse Gelsinger," *New York Times Magazine*, 11/28/99

6 Rick Weiss and Deborah Nelson, "Penn Settles Gene Therapy Suit," *Washington Post*, 11/04/00

7 Huntly Collins, "Penn Team Finds Clue to Gene-Drug Death," *Philadelphia Inquirer*, 01/26/01

8 Weiss and Nelson, 11/04/00, op. cit.

9 Marie McCullough, "Lawyer Sees His Role As Warning to Clinical Researchers," *Philadelphia Inquirer*, 05/20/02

10 Weiss and Nelson, 11/04/00, op. cit.

11 "Gene Therapy Death Lawsuit Settled," CBS, 11/03/00

12 Kristen Philipkoski, "Perils of Gene Experimentation," *Wired*, 02/21/03

13 Rick Weiss and Deborah Nelson, "Requests Spur Debate About Openness," *Washington Post*, 10/30/99

14 Deborah Nelson and Rick Weiss, "NIH Not Told of Deaths in Gene Studies; Researchers, Companies Kept Agency in the Dark," *Washington Post*, 11/03/99

15 Deborah Nelson and Rick Weiss, "Gene Test Deaths Not Reported Promptly," *Washington Post*, 01/31/00

16 Jackie Smith, "Trust us: We're Medical Researchers," *Toronto Globe and Mail*, 01/14/03

17 Linda Pannozzo, "The Race to Own the Body: Cashing in on the Human Genome Project," *HighGrader Magazine*, May/June 2000

18 ibid.

19 *FDA Consumer*, op. cit.

20 Nelson and Weiss, 11/03/99, op. cit.

21 Nelson and Weiss, 01/31/00, op. cit.

22 Henry David Thoreau, *Journal*, 11/11/1850, published 1903

23 See www.scid.net for links.

24 Reuters, 04/28/00, citing *Science*. A rather full description can be found in the transcript of a February 2003 FDA meeting, at www.fda.gov/ohrms/dockets/ac/03/transcripts/3924T2_01.htm

25 The numbers are from *Nature*, 06/10/04. There have been slight variations: *The Scientist*, 10/20/03, said ten at Necker and four at Great Ormond Street; the American Society of Gene Therapy (ASGT) in a 10/03/02 press release said ten of eleven patients at Necker had been successfully treated.

26 It was suggested that the first case might have been connected with a case of chicken pox that could have been an additional trigger for the leukemia, but even before the second case the FDA's Philip Noguchi said: "The evidence that the treatment helped trigger the cancer is definitive." Rick Weiss, "Resumption of Gene Therapy Urged; Advisory Panel Backs Restrictions After Treatment Triggered French Boy's Illness," *Washington Post*, 10/11/02

27 Jo Lyford, "Gene Therapy 'Caused T-cell Leukemia,'" *The Scientist* 10/20/03, www.biomedcentral.com/news/20031020/02/; citing *Science*, 302:415-419, 10/17/03

28 "FDA Places Temporary Halt on Gene Therapy Trials Using Retroviral Vectors in Blood Stem Cells," *FDA Talk Paper*, 01/14/03

29 Andrew Pollack, "Gene Therapy Trials Halted," *New York Times*, 01/15/03

30 "Gene therapists hopeful as trials resume with childhood disease," *Nature*, 06/10/04

31 "Gene therapy put on hold as third child develops cancer," *Nature*, 02/09/05

32 Jackson J. Spielvogel, *Western Civilization*, 5th edition, Wadsworth, Belmont, CA, 2003, p. 149

33 Dan Hurley, "With New Science, Hair Restoration Improves," *New York Times*, 06/15/04; www.hairlosstalk.com

34 "Researchers Identify Signals that Cause Hair Follicles to Sprout," Howard Hughes Medical Institute, 03/20/03; *Nature*, 03/20/03

35 "Stem cell tricks hint at baldness cures," *New Scientist*, 03/14/04

36 Helen Pilcher, "Hair-Raising Stem Cells Confirmed in Mouse Skin," *Nature* online, 09/02/04

37 Robert Preidt, "Stem Cells Found in Hair Follicles," *Forbes* online, 09/02/04

38 "Single Isolated Mouse Skin Cell Can Generate into Variety of Epidermal Tissues," *Science Daily*, 09/06/04

39 American Society of Plastic Surgeons, "2000/2001/2002/2003 National Plastic Surgery Statistics"

40 Bill McKibben, "Designer Genes," *Orion*, 05–06/03

41 Andrew Kimbrell, *The Human Body Shop*, Washington, D.C.: Regnery Publishing, 1997, pp. 167–185

42 "A Hormone to Help Youths Grow Is Approved by F.D.A.," AP, in the *New York Times* 07/27/03

43 Bruce Downie et al., "Are Short Normal Children at a Disadvantage?" *British Medical Journal*, 01/11/97

44 Kimbrell, op. cit., p. 176

45 "Safer Route to Gene Therapy Found," BBC, 09/22/04

46 Pollack, 09/14/04, op. cit.

47 Claudia Dreifus, "A Doctor Puts the Drug Industry Under a Microscope," *New York Times*, 09/14/04

48 Dr. Aubrey de Grey, in Audrey Gillan, "Psst . . . The Secret of Youth Can Be Yours for £250," *Guardian*, 09/11/04

49 "First Parkinson's Gene Therapy Patient Well After 1 Year Suggesting That the Therapy Is Safe And Effective," www.news-medical.net, 09/02/04

50 "Alzheimer's Gene Therapy Drug Developed," Betterhumans, 09/21/04

51 Ceregene, Inc., Press Release, 09/21/04

52 "New Gene Therapy Technique May Help Treat Alzheimer's Disease," www.news-medical.net, 09/21/04

53 Lindsey Tanner, "Walking Routine Can Help Fight Alzheimer's," AP, in the *Chicago Sun-Times*, 09/22/04

54 "A Nice Cup of Tea Could Hold Back Alzheimer's, Scientists Say," AFP, 10/26/04

55 Ruth Hubbard and Elijah Wald, *Exploding the Gene Myth*, Boston: Beacon Press, 1993, 1999

56 The Council for Responsible Genetics, "Position Paper on Human Germline Manipulation," updated Fall 2000

57 Roger Highfield, "'Designer Sperm' Can Weed Out Gene Defects," London *Daily Telegraph*, 10/23/01

58 BBC, 01/20/03

59 www.frenchanderson.org/inutero/pdf/rac.PDF. The minutes of the RAC meeting are at www4.od.nih.gov/oba/rac/minutes/9-98rac.htm.

60 www.genetics-and-society.org/analysis/promodeveloping/anderson.html

61 www4.od.nih.gov/oba/rac/racinutero.htm

62 United Nations Development Program, *Human Development Report 2004*, New York, 2004, p. 156

63 ibid., p. 168

64 ibid.

65 "Bespoke Medicine," *New Scientist*, 09/11/04

66 "Royal Society to Investigate Potential of Personalised Medicines," Royal Society Press Release, 09/20/04

7

GENE DOPING IN SPORTS

■ HEALTHY VOLUNTEERS ■

APPROPRIATE CANDIDATES for experimental, and potentially dangerous, new procedures can be hard to find, as examples from **Chapter 6** demonstrate:

▶ Terminally ill patients who are competent enough to give consent are ethically justifiable subjects; but if they get sick, how do you tell if this was because of the experiment?

▶ Testing on babies too young to give consent is a more complicated decision, since the parents themselves may be so emotionally entangled that they have trouble making good decisions; X-SCID babies with no tissue-matching donor may be a clear-cut case, but this is an unusual situation.

▶ Bioethicist Arthur Caplan recommended testing, where possible, on competent adults because "you can't use people who can't consent—the mentally ill, babies, retarded people," but Henry I. Miller counters that you should test "on people who are unlikely to be worse off"—and Jesse Gelsinger, who was chosen under Caplan's guidelines, died.[1]

Gelsinger is only the starkest reminder of the risks of testing on healthy (or relatively healthy) people. And yet there is a group of people who are willing, indeed eager, to try anything that promises to enhance their natural abilities—athletes.

■ DRUG ABUSE BY ATHLETES ■

DR. BOB GOLDMAN, founder of the US National Academy of Sports Medicine, has been researching athletes' attitudes to performance enhancing drugs since 1983. They haven't changed significantly. His first survey covered 198 Olympic-class athletes, from various disciplines. He asked them two hypothetical questions:[2]

▶ If you were offered a drug that guaranteed both that you will win and that you won't get caught, would you take it? 195 of 198 (98 percent) said yes.

▶ If the drug had side effects so that you were certain to win for five years but then you would die, would you take it? *103 of 198 (52 percent) said yes.*

If Goldman's sample is typical—and it certainly seems to be, from his subsequent surveys—then more than half of all world-class athletes will accept not just a risk but the *certainty* of death, if they get success first. Some of the people he has asked were only sixteen years old. As he says, "To be willing to die at 21 is a serious psychological mindset that must be addressed."

The short-term thinking of highly motivated athletes makes a mockery of informed consent (see **Box 7.1**). And it's not just an issue at the professional level. In the same week that Professor Sweeney (see below) announced that he had genetically engineered mice to be abnormally muscular, he got a call from a high school football coach, who wanted to pump up his whole team.[3] Then a wrestling coach. And the calls kept coming, every week for months, mostly from wrestlers and weightlifters.

HUMAN GENETIC ENGINEERING

7.1 LYLE ALZADO

LYLE ALZADO WAS an NFL star who died on May 14, 1992, at the age of forty-three, from a rare brain cancer, and tried to make himself into a cautionary tale.

Alzado was small for a football player. He began taking steroids in college and bulked up to 254 pounds.[4] (In 1981, there were only 23 professional players who weighed more than 300 pounds; by 1991, there were 370; in 2003, 321.[5]) He played fifteen seasons in the National Football League (NFL), and spent about $30,000 a year on steroids and, later, human growth hormone. This may have looked like a sound investment at the time, to those who consider such decisions from a financial perspective, but the addictive thrill of competition was a more important motive.[6] "I was so wild about winning. It's all I cared about, winning, winning."

His second wife (of four) blamed the failure of their marriage on his "'roid rage," and Lyle became convinced that the drugs caused his cancer, though no link has been proved. He came clean in Sports Illustrated, in a July 1991 cover story titled "I'm Sick and I'm Scared," and urged young people, "It wasn't worth it. If you're on steroids or human growth hormone, stop. I should have."

There is no evidence that they listened. Most young athletes have never heard of Lyle Alzado.

It's possible that the major league professionals didn't call Professor Sweeney because they know enough to stay away from such preliminary and experimental procedures. Or perhaps they just didn't trust him to keep the call confidential.

■ THE BALCO SCANDAL ■

A NORTHERN CALIFORNIA company, BALCO, which officially dealt in legal dietary supplements, was the subject of widely publicized accusations in 2003–2004 that it had developed a kind of steroid deliberately designed to evade drug tests.[7]

■ 160 ■

The point about this substance was not that it had new physiological benefits, simply that it was hard to detect.

A grand jury has been investigating the BALCO affair, and most of what is thought to be known about it (in early 2005) either comes from leaks or is the subject of litigation, or both. Some very high profile athletes have been accused of taking drugs supplied, directly or indirectly, by BALCO, including:[8]

- Olympic gold medal sprinter **Marion Jones** (who denies it and is suing BALCO's founder, Victor Conte, for defamation)
- Jones's second husband, 100-meter world record holder **Tim Montgomery** (who denies it)
- Baseball superstar **Barry Bonds** (who denies knowing of it but is reported to have admitted he may have been given steroids in a cream)
- Baseball star **Jason Giambi** (who denied it publicly but is reported to have admitted it to the grand jury)
- Sprinter **Kelli White** (who first denied it and then admitted it, cooperating with authorities reportedly in exchange for a shorter suspension)
- British sprinter **Dwain Chambers** (banned for life from the Olympics by UK authorities)

Many more track-and-field athletes, at least seven of whom have been suspended by the sport's governing authority, and several less famous baseball players have also been involved in the scandal. This large web of athletes seems to be connected by word of mouth, which is understandable given the illegal nature of the substances involved. It appears that they got away with it for several years.

The scandal was discovered only because Jones's ex-husband, himself a former steroid user, gave her ex-coach a syringe containing a mysterious liquid. The coach, Trevor Graham, sent the syringe anonymously to doping authorities. They analyzed the contents, identified them as a previously unknown steroid, and engineered a new test that could detect it.

There have been other reports, possibly exaggerated, that a large proportion of professional athletes are or have been users of performance enhancing drugs. Conte himself, in the TV interview during which he accused Jones, said, "My guess is that more than 80 percent [of baseball players] are taking some sort of a stimulant before each and every game."[9] He may be exaggerating: "Everybody's doing it" is not a legal justification, but it might help his case. Still, it does seem that the use of drugs in sports is widespread. Even the governor of California, Arnold Schwarzenegger, used them in his bodybuilding days, although he notes that they were not illegal until 1990, and he denies having used them since then.

Steroids have side effects: Kelli White has described having periods every two weeks that lasted for a week, developing acne, and having a notably raspy voice; she also gained fifteen pounds of muscle and won two gold medals at the World Championships. The drugs worked.

Even if gene doping is dangerous, there will be people willing to take that risk, if it works. But will it? It might.

■ GENES AND ATHLETES ■

IT IS OFTEN said that top-class athletes are "genetic freaks," and there is some truth to that. Richard Pound, the outspoken head of the World Anti-Doping Agency (WADA; see **Box 7.2** and **Box 7.3**), has said: "There is some genetic selection anyway. If you are a five-foot-two-inch Malaysian, you are probably not going to be a basketball star. And if you are a seven-foot Chinese you are not going to be a very good badminton player. If two basketball players marry, the chances are they will have tall children. I don't think we are at the stage where a young couple says 'What will we have this time, dear? A gymnast or a rower?'"[10]

7.2 THE WORLD ANTI-DOPING AGENCY (WADA)

WADA WAS FOUNDED in 1999 to eradicate drugs from sport and standardize the rules all over the world and in every discipline.[11] An independent foundation, originally set up by the International Olympic Committee (IOC), it is now financed half by the IOC and half by national governments. By fall 2004, 153 countries had signed the "Copenhagen Declaration" recognizing the role of WADA and the Code it has developed (**Box 7.3**).

Under the leadership of Richard (Dick) Pound, a former Olympic swimmer (twice a finalist in 1960) who has drawn criticism for his abrasive style but has certainly established the organization, WADA has signed up virtually every international sports federation. It has developed new and more effective drug tests and moved swiftly to implement them.[12]

One problem is that the National Football League (NFL), Major League Baseball (MLB), and the Nation al Basketball Association (NBA), among others, have not signed up. The owners cite their collective bargaining agreements with the players' unions as preventing them from enforcing the standards that WADA demands.[13] As these agreements come up for renewal, WADA hopes to bring them up to code.

7.3 THE WADA ANTI-DOPING CODE

THE RULES EXPRESSED in the WADA Code are both strict and general. They begin with this clear statement of intent,[14] "The use of any drug should be limited to medically justified indications."

The Code lists all known banned substances, carefully including phrases such as "including other substances with a similar chemical structure or similar biological effect(s)." There is no "wiggle room." Essentially, if a drug is in your body then you flunk, even if you did not intend to take it, and if you miss a test you are assumed to have failed it.

Gene doping is high on the WADA agenda, and has been since at least 2002, when WADA held a weekend conference on Genetic Enhancement of Athletic Performance at Cold Spring Harbor Laboratory on Long Island.[15] Partly as a result of that event, gene doping was added to the Code, under the section "Prohibited Methods." For 2005, the relevant part reads:

7.3

M.3 Gene Doping

The non-therapeutic use of cells, genes, genetic elements, or of the modulation of gene expression, having the capacity to enhance athletic performance, is prohibited.

Note that all forms of tampering at the genetic level are banned. Gene transfer is only one of the possibilities.

The difficulty is detection. Muscle biopsies, the most certain form of testing, are considered "too invasive" even by Pound.[16] There is as yet no definite alternative, and there seems to be a race between the cheats and the authorities. WADA is spending one-fifth of its research budget to find a test; but that is only $1 million a year.[17] If someone can develop a test before gene dopers can develop a practical method of cheating, it will be a considerable achievement.

Most of us will never complete a marathon, let alone do so in close to two hours. Nor will we ever run 100 meters in less than ten seconds. We will not reach world class, no matter how hard we try, even if we take every drug known at present—testosterone, erythropoietin (EPO), human growth hormone (HGH), steroids of all kinds, vitamins from A to Z, everything—and top them off with "blood doping" (last-minute transfusions of stored red blood cells, our own or someone else's), caffeine, cocaine, and amphetamines.

We don't have the genes. Actually, it's not completely clear what "the genes" for maximum athletic performance are, given the complexity of the system. Attempts are being made to find out: Australia has set up a national program to collect DNA samples from exceptional athletes, with the idea of identifying future superstars.[18] (They take sports seriously in the Antipodes: Despite a population of only 20 million, Australia came fourth overall in the 2004 Olympic medal table, with 49; the US, with almost fifteen times the population, had 103.) Scientists at the Australian Institute of Sport and the Sydney Royal Prince Alfred Department of Molecular and Clinical Genetics are already reported to have identified two performance genes that provide power and stamina.

That, and similar work, might help us to pick out potential talents. But what about *making* them?

■ GENE DOPING? ■

To JUDGE BY some of the hype bandied about in 2004—which, by no coincidence, was an Olympic year—the era of gene doping had already arrived.

Theodore Friedmann, director of the Program in Human Gene Therapy at the University of California at San Diego (who may, perhaps, be inclined to the view that gene therapy can work) was quoted as saying,[19] "All the technology is in the medical literature. The genes are all available or you can make them. The vectors, the viral tools, are all published and available. All it takes is three or four reasonably well-trained postdocs and a million or two dollars."

That is, you could use gene transfer (obviously, in this application, it is not "therapy") to grow muscle and to increase the production of red blood cells—to affect both strength and endurance. Or, as Dr. Peter Weygand, of Rice University, opined, with perhaps just the merest hint of overstatement,[20] "You don't need to lift weights, and you don't need to go on 10-mile runs to train for endurance. It [genetic modification] would replace training; it would make training seem trivial and more than obsolete. Somebody who's not athletic at all could be transformed into something superhuman."

In fact, some experts are complaining that media reports are giving athletes a "bluffer's guide to cheating."[21] Professor Geoff Goldspink, of University College, London, has formally complained to WADA about the level of detail published, though it is hard to see how traditional scientific discourse can effectively be combined with secrecy—or what WADA could do about it. A more sober, and likely more realistic, view was expressed by Professor Lee Sweeney of the University of Pennsylvania, who is at the forefront of the experiments whose publication helped to cause the 2004 fuss:[22] "No one should worry about (genetically altered) athletes today, but if you're WADA, you've got to anticipate so you won't be blindsided when that day comes." Certainly, recent experiments on mice, and to some extent other animals, do suggest that there is something to worry about.

The research that could be abused for gene doping, however, does have potentially important medical applications. Professor Sweeney and his colleagues (including Dr. Nadia Rosenthal and others at Harvard) are hoping to reverse the loss of muscle that occurs either normally, as part of the aging process, or as a result of muscular dystrophy.[23] They are aiming at rebuilding the various types of muscle tissue (see **Box 7.4**), but athletes are *not* the focus or intended targets.

BOX 7.4 SPRINTERS AND MARATHONERS

WE ALL HAVE different kinds of muscles. One way of categorizing them that is especially relevant to athletes is:

- **Type I**, or **"slow-twitch,"** muscles are **aerobic**—that is, use oxygen— and get their energy by metabolizing fats; they can keep going for a long time, but are not good for short, intense activities like lifting heavy objects.
- **Type II**, or **"fast-twitch,"** muscles are **anaerobic**, burn sugar fast— which does not require oxygen—and get tired very quickly, but are ready for action at a moment's notice.

Many of us have about half-and-half. Athletes who specialize in endurance (any long distance running, swimming, cycling, or similar event) have up to 90 percent type I or "slow-twitch" muscles. Sprinters, weight lifters and all those involved in explosive types of event have perhaps 80 percent type II.

7-4   Some of the differences between people are innate, others are affected by training. If you want to run long distances, for example, you practice that, and increase your proportion of slow-twitch muscles. Similarly, sprinters work on developing their fast-twitch muscles. Different types of gene doping might be needed for each, and both may be on the horizon.

However, their experiments, and others mentioned below, are so obviously applicable to sports—and open to abuse—that Sweeney has become a prominent spokesman *against* gene doping, just as Ian Wilmut, leader of the team that cloned the first mammal (see **Chapter 3**) became a prominent spokesman against the reproductive cloning of people.[24]

Both Sweeney and Wilmut are in the unfortunate position of enabling technologies of which they disapprove. So, in a sense, was Albert Einstein. But Sweeney's position is particularly difficult: Einstein's work was deeply theoretical; Wilmut hoped to improve the production of GE medicines; but Sweeney and his colleagues are trying to make something that will directly help sick people—and, unchanged, can be abused. It's a hard place to be.

■ MIGHTY MICE AND RATS ■

MUSCLE GROWTH IS (to simplify) in part a response to insulin growth factor I (IGF-I), a protein produced by a specific gene. Researchers injected a synthetic form of the IGF-I gene into the muscles of young mice. This new gene did not have to be in every cell of the body, just the ones they hoped to affect. Muscle cells can last a long time, and it is possible that one "fix" would last an older person for the rest of their life, since that gene will continue to produce the protein until the cell dies. Injecting the protein itself would induce a temporary effect, but that would wear off too soon to be of much use.

With no exercise, the muscles grew 15–30 percent more than

normal. Older mice maintained their muscle strength, and sedentary rats also gained about 15 percent in strength. They got stronger without *doing* anything. Then the researchers put rats with one leg "enhanced" and the other normal on a weight training regimen. (The things they do to those animals boggle the mind.) The injected muscles gained nearly *twice as much* strength as the uninjected ones, and lost it more slowly when the training ended. The effect seemed to last for the lifetime of the rats.

In good news for patients, but bad news for dope testers, the experiment was designed so that the IGF-I gene "stays in the target muscle and does not move into the bloodstream where it could cause damage to other organs."[25]

■ MYOSTATIN IN MICE AND BULLS AND PEOPLE ■

ANOTHER PART OF the physiological equation is performed by a substance called **myostatin**, which *inhibits* muscle growth and is also involved in the development of fat. So inhibiting that inhibitor would theoretically promote large, lean muscle. And indeed, researchers have done exactly that to mice. First they succeeded in genetically engineering mice not to have the necessary gene, then they engineered mice to have additional proteins that block the effect of the gene, rather than simply deleting it.[26] It worked.

It was already known that myostatin inhibition also works in animals other than mice. A breed of cattle called the Belgian Blues has a genetic mutation that produces an ineffective version of myostatin, and it is exceptionally lean and over-muscled. But would it work in people?

No one has tried altering a human experimentally, but the June 2004 *New England Journal of Medicine* described a baby that had a similar mutation.[27] The boy's upper arm and upper leg muscles were twice as large as normal, which led doctors to investigate. At five, he was still "much stronger" than his contemporaries, but seemed healthy. There are also reports that a European weight lifter had the same genetic combination.[28]

Gene transfer, to put the "low-myostatin" gene into an athlete's muscles, is a long way off and may actually never be necessary. There are already "protein supplements" on the market that claim to let you "bust right through your genetic growth ceiling" by taking a "natural myostatin neutralizer" that "can help supercharge growth to an entirely new level!"[29] (Caveat emptor!) More conventional drug companies are working on pills that would have this effect—for medical, not enhancement, reasons. The use of these would, of course, be outlawed in sports by WADA; in fact, it already is (see **Box 7.3**).

■ MARATHON MICE ■

Two DIFFERENT TEAMS of scientists have used two different methods to genetically engineer mice so that they have greatly increased endurance. They both work to encourage aerobic, rather than anaerobic, muscle activity (see **Box 7.4**).

One experiment, conducted by Professor Ronald Evans and others at the Salk Institute in San Diego, essentially switched to permanently "on" a gene that produces a protein, PPARδ, that tells muscles to burn fat instead of sugar—which requires oxygen and is therefore an aerobic process.[30] Fat burning was the researchers' interest, so they fed the GE mice, and a control group, a high fat diet for ninety-seven days. Sure enough, the GE mice burned more fat, and gained only one-third of the weight that their normal counterparts did.

It also turned out, however, that the GE mice developed a higher proportion of slow-twitch muscles, just as if they had been in marathon training. And the "training" worked: The mice were put on a treadmill and made to run to exhaustion. Standard mice managed ninety minutes and covered 2,950 feet; the GE mice kept going for another hour and covered 5,900 feet.[31]

This is just one finding, one contribution to research on muscles—and it's less than no use at all to anyone who wants to be the "World's Fastest" 100-meter runner. There is a lot left to learn, and the summary for the non-scientist provided by the journal in which the study was published appropriately offers this warning:[32] "Maybe Olympians can be made after all—but don't give up on training just yet. A full understanding of the molecular basis of muscle fiber determination, including the interactions between PPARδ and its regulatory components, awaits further study."

The other experiment, which interfered with the ability of muscles to switch from aerobic to anaerobic activity, produced some strange results. The GE mice did better going uphill than their normal counterparts, but worse going downhill. That suggests some interesting avenues of research, not to mention confirming the truism among runners that there is such a creature as a "downhill racer," but there was a serious problem: After four days, the GE ones were plumb worn out,[33] "The mutants' run time was significantly shorter and their muscles were clearly damaged."

As Professor Randall Johnson, one of the study's authors, wryly noted, "I'm not going out and buying running shoes today."[34]

■ DEAD MONKEYS ■

YET ANOTHER EXPERIMENT highlights the unpredictable possibilities of serious, even fatal, harm. Dr. James Wilson, a long-standing gene therapy pioneer who was forced to move from human to animal studies after the tragic death of Jesse Gelsinger (see **Box 6.2**), went on to work on ways of introducing single genes into muscle cells. In particular, he tried to introduce a gene that increases the production of red blood cells. This would be like the use of EPO boosters by current athletes, except that it is done on a cellular level. It worked all too well.[35]

Within three weeks, the eight monkeys he injected had red blood levels higher than the most extreme athletic dopers. Within a month, the blood was so thick it threatened to kill them. The researchers injected substances to thin the blood, and four of the monkeys were kept alive that way for a year. The other four, however, began completely rejecting not only the EPO produced by the gene that had been introduced but also the normal, baseline EPO their bodies needed. They got severe anemia, and had to be killed. Wilson was "perplexed."

This is good science, although (in a sense, *because*) the experiment failed; it produced interesting data. It is, however, bad therapy—in fact, not therapy at all. As Wilson himself said,[36] "It's hard for me to believe that [any athlete] would try to do this."

■ LEGALIZE IT? ■

SOME PEOPLE DO suggest that we should simply accept gene doping, and perhaps regulate it by sorting athletes into classes. In fact we already do that, in a sense—we select by chromosomes and, in most sports, men and women compete separately. (Equestrianism is the only exception among Olympic sports, presumably because the horse does so much of the work.) As the British bioethicist Andy Miah has pointed out, the paralympics is also based on a concept of sorting, and so in a sense are boxing and weight lifting, which group competitors by weight.[37]

The long-term consequences of steroid use are less certain than the worst scare stories suggest, because long-term studies are hard to do. Lyle Alzado (see **Box 7.1**), may have been wrong in attributing his brain cancer to steroid abuse; that was never proved. Nor was there any direct connection between the 2004 death of former baseball star Ken Caminiti and his long-term abuse of both steroids and other illegal drugs.[38] It was suggested by some commentators that the emotional consequences of *withdrawing* from steroids led Caminiti to compensate by using cocaine, which seems plausible though anecdotal. At any rate, that steroids are harmful is essentially beyond dispute.

Many of the arguments in favor of allowing drugs and genetic interventions and other such aids to performance assume that they work predictably, reliably, and safely (which is often not true). Thus Professors Julian Savulescu and Bennett Foddy, of the University of Oxford and Murdoch Children's Research Institute, write in a paper called "Why We Should Allow Drugs in the Olympics":[39] "We should permit drugs that are safe, and continue to ban and monitor drugs that are unsafe."

That sounds reasonable until you think about it in the context they are using. Savulescu and Foddy claim that taking a moderate dose of EPO is "not a problem" and "allows athletes to correct for natural inequality." (Excuse me?) But do they really imagine that a human athlete, told that something will benefit them, will voluntarily restrict themselves to taking just a little bit? Is that plausible? And if they mean that oxygen carrying capacity may, when tested, be so high and no higher, how on earth does this differ in principle from the regime we have now?

We do encourage athletes to eat healthy food, and even to take vitamins up to a point. However, it takes a special kind of mind not to be able to tell the difference between eating healthy food and undergoing extreme and possibly dangerous pseudo-medical procedures. Most ordinary people have no trouble making that distinction.

It is true that some kinds of performance enhancers are allowed—"high-tech swimsuits, oxygen deprivation tents, even Gatorade."[40] When athletes are injured, we expect them to get medical treatment; and there could theoretically come a point—perhaps—where gene therapy might be useful in healing. Would that be allowed? Could an athlete use an injected protein, with limited applicability and life span, rather than a gene that would remain longer? We are far away from that, but there is a discussion to be had, and there will continue to be lines that need drawing. Certainly that theoretical scenario is no reason to allow anything that might happen before it's even possible.

Professor John Hoberman of the University of Texas at Austin has studied sports doping over the years and is in no doubt about what legalizing genetic modification would lead to:[41] "It's a terrible situation. . . . it's going to turn [sports] into a kind of circus—a freak show." And then there are the side effects. The monkeys *died*. Some of the mice wore out and damaged their "enhanced" muscles. No one knows what strains excessive muscle development would place on bones and tendons, as Dr. Peter Weyand, who studies muscles and movement at Rice University, says:[42] "All bets are off when you start playing with genetic engineering . . . in terms of system function, organ function, and long-term effects. . . . There might be deleterious health effects. We really don't know."

Legalizing gene doping, or indeed drug doping, is one of those libertarian notions (see **Chapter 10**) that makes for provocative publicity, and not much else. Every reputable governmental body everywhere, and every sports organization, is against it. It's not going to happen.

■ FURTHER READING ■

Free Documents from the Web

Stuart Stevens, "Drug Test," *Outside*, 11/03, is a fascinating and very detailed account of self-medicating by a talented amateur athlete, at http://outside.away.com/outside/bodywork/200311/200311_drug_test_1.html

The World Anti-Doping Agency (WADA) Website, www.wada-ama.org, includes the regularly updated "Prohibited List: World Anti-Doping Code," fact sheets and other reports.

Winning at Any Cost: Doping in Olympic Sports, by the CASA National Commission on Sports and Substance Abuse, is a major, official Report (113 pages, 1.8mb as a pdf), funded in part by the U.S. Office of National Drug Control Policy, available at www.casacolumbia.org/pdshopprov/files/34968.pdf.

Diane Kightlinger, "Citius, Altius, Fortius—Purius? Doping and Olympic Athletes," *Update*, the magazine of the New York Academy of Sciences, 08–09/04, is at www.nyas.org/snc/update.asp?updateID=8&page=1.

Sal Ruibal, "Tackling Longtime Issue of Drugs No. 2 on Sports Changes Wish List," *USA Today*, 09/09/04; www.usatoday.com/sports/2004-09-09-ten-changes-drugs-testing_x.htm

Tim Radford, "Gene Cheats: The New Risk Posed to World Sport," *London Guardian*, 02/17/04; http://sport.guardian.co.uk/news/story/0,10488,1149824,00.html

Gregory M. Lamb, "Will Gene-Altered Athletes Kill Sport?" *Christian Science Monitor*, 08/23/04; csmonitor.com/2004/0823/p12s01-stgn.html

■ ENDNOTES ■

1 Arthur Allen, "Bioethics Comes of Age," Salon.com, 09/28/00

2 Ellie Tzortzi, "Should the X-men Have a Go at the Games?" Reuters, 08/29/04

3 Alan Zarembo, "DNA May Soon Be in Play," *Los Angeles Times*, 08/27/04

4 Mike Puma, "Not the Size of the Dog in the Fight," ESPN.com special

5 "Morituri Te Salutamus," *Economist*, 01/18/92

6 S. Smith, "I'm Sick and I'm Scared," *Sports Illustrated*, 07/08/91

7 The *San Jose Mercury News* published a useful timeline about the development of the scandal (06/03–07/04) at www.mercurynews.com.

8 The *San Francisco Chronicle* has broken many of these stories, and covered them all in articles in late 2004, notably on 12/16/04, 12/13/04, 12/02/04, 08/23/04 and 07/31/04.

9 *San Francisco Chronicle*, 12/03/04

10 Tim Radford, "Gene Cheats: The New Risk Posed to World Sport," *Guardian*, 02/17/04

11 See www.wada-ama.org

12 "Ever farther, ever faster, ever higher?" *The Economist*, 08/05/04

13 Sal Ruibal, "Tackling Longtime Issue of Drugs No. 2 on Sports Changes Wish List," *USA Today*, 09/09/04

14 World Anti-Doping Agency (WADA), "The 2005 Prohibited List: World Anti-Doping Code," issued 09/23/04, to come into force 01/01/05

15 News Release, "WADA Conference Sheds Light on the Potential of Gene Doping," 03/20/02

16 *Economist*, 08/05/04, op. cit.

17 Zarembo, 08/27/04, op. cit.

18 "Designer Athletes from DNA Tech?" AFP, 08/30/04

19 Diane Kightlinger, "Citius, Altius, Fortius—Purius? Doping and Olympic Athletes," *Update*, 08–09/04

20 Gregory M. Lamb, "Will Gene-Altered Athletes Kill Sport?" *Christian Science Monitor*, 08/23/04

21 Peta Bee, "Sport Braced as the Gene Genie Escapes from its Bottle," *Guardian*, 09/13/04

22 Sal Ruibal, "The Fight Against Gene Doping," *USA Today*, 04/14/04

23 H. Lee Sweeney, "Gene Doping," *Scientific American*, 07/04

24 Rudolf Jaenisch and Ian Wilmut, "Don't Clone Humans!," *Science* 2001 291: 2552

25 Paul Recer, "Gene Therapy Creates Super-Muscles," AP, 02/16/04

26 "New 'Mighty Mice' Research Brings Muscle Growth Closer To Reality," Johns Hopkins Press Release, 07/16/01, announcing a publication in *Proceedings of the National Academy of Sciences*, 07/16/01

27 "Secrets of Muscle Growth Revealed," BBC, 06/24/04

28 Nicholas D. Kristoff, "Building Better Bodies," *New York Times*, 08/25/04

29 "Pinnacle Juiced Protein," www.nicemuscle.com

30 Helen Pearson, "Geneticists Engineer Marathon Mice," *Nature*, 08/23/04

31 "Gene Targeting Turns Mice into Long-Distance Runners," Public Library of Science, 08/24/04

32 ibid.

33 "A Skeletal Muscle Protein That Regulates Endurance," Public Library of Science, 08/24/04

34 Pearson, 08/23/04, op. cit.

35 Zarembo, 08/27/04, op. cit.

36 ibid.

37 Lamb, 08/23/04, op. cit.

38 " '96 MVP Admitted Steroid Use, Fought Drug Problem," ESPN, 10/11/04

39 Professors Julian Savulescu and Bennett Foddy, "Why We Should Allow Drugs in the Olympics," www.practicalethics.ox.ac.uk

40 Lincoln Allison, "Faster, Stronger, Higher," *Guardian*, 08/09/04

41 Lamb, 08/23/04, op. cit.

42 ibid.

8

EUGENICS: LEARNING FROM THE PAST

■

■

■ INTRODUCTION ■

EUGENICS HAS BECOME a term of abuse, and for very good reason. It represents an idea that caused misery in the twentieth century, to an extent that is almost incredible.

In Nazi Germany, it led to genocide.

But eugenics was not only advocated by fascists and their supporters. It was always essentially about attempts to "improve" the human race, and this is a goal that over the years has been adopted not only by vicious racists but also by well-meaning progressives, as well as social conservatives and intellectuals from all across the political spectrum.

If this were just history, it would still be important to remember, to pass on, and to learn from. But it's not. Far from it. It's coming back.

If we are to avoid tragedies, both similar to and different from those of the past, it is important that we see eugenics for what it was—an appalling but originally well-intentioned idea—and we must understand how closely the modern advocates of human genetic engineering follow its fundamental concepts.

■ STILL "IMPROVING" SOCIETY? ■

MILLIONS OF PEOPLE have been killed in the name of eugenics, and millions more sterilized. After 1945, the concept was indeed discredited for many years, but it never really went away. Indeed, many American states continued to enforce sterilization laws for decades after that. (It was only in 2002 that

North Carolina finally repealed its law, and the governor, as well as those of Virgina and Oregon, finally made formal apologies to survivors and their families.[1])

In 1974, US District Court Judge Gerhard Gesell revealed that:[2] "Over the last few years, an estimated 100,000 to 150,000 low-income persons have been sterilized annually in federally funded programs" in which they were "improperly coerced into accepting a sterilization operation under the threat that various federally funded benefits would be withdrawn."

Compulsion can take various forms. In India in the late 1960s, bribery was used—men were offered a transistor radio in exchange for having a vasectomy. In Louisiana, as recently as February 2000, a district judge offered a defendant the choice, if it can be called that, between sterilization and imprisonment:[3] "I don't want to have to lock you up to keep you from having any more children, so some kind of medical procedure is needed to make sure you don't."

The judge was clearly speaking in what he regarded as the best interests of the woman involved, as well as the community as a whole. (She had eight children, and was convicted of abusing three of them.) That is the whole point. Eugenics was not dreamed up as a tool of hatred and racism, but as a *benevolent* force for the *improvement* of society.

In the 1920s and 1930s, eugenics was mainstream thinking. A lot of very smart and well-meaning people were fooled. That is precisely what should give us in the early twenty-first century pause, as we face attempts to rehabilitate eugenics in a slightly different form.

■ REVIVING EUGENICS ■

THE RESPECTABLE ADVOCATES of Human GE all condemn the excesses of twentieth-century eugenics. (There may be a neo-

Nazi fringe of whom that is not true.) Indeed, they mostly shy away from the word itself, even though they are in fact effectively reviving the concept.

- Almost every recommendation for "germline gene therapy" is accompanied by a denial that it constitutes eugenics.
- Virtually every discussion of the Human Genome Project includes a repudiation of eugenics.
- The site of the former Eugenics Record Office, now the Cold Spring Harbor Laboratory, hosts a powerful denunciation of its own past, built under the leadership of James Watson, who is also one of the most passionate advocates of germline intervention.[4]

There is, however, a burgeoning movement to revive eugenics, quite specifically. For example:

- Richard Lynn, professor emeritus of psychology at the University of Ulster, wrote a book called *Eugenics: A Reassessment*.[5] He then co-authored an even more controversial book called *IQ and the Wealth of Nations*, which purported to demonstrate that Africans are poor because they are stupid.[6]
- The late Glayde Whitney, a professor at Florida State, promoted eugenics in his foreword to the autobiography of former Klansman David Duke.[7]
- A September 2004 Conference at the Royal Society in London included sessions on choosing our children's sexual orientation and preventing the existence of people with disabilities and placed this in the context of learning from the history of eugenics.[8]

There are numerous websites and sympathizers (see **Chapter 10**). The out-and-out racists attract some attention periodically, but are clearly fringe characters. Of far more concern, because the prejudices involved are more subtle, are those who—like Watson—absolutely disavow old-fashioned state-enforced eugenics but promote a privatized version.

David Galton (no relation to the man who coined the term; see below) addresses the issue directly in his 2001 book *In Our Own Image: Eugenics and the Genetic Modification of People*.[9] He condemns both the Nazi application and the less extreme versions found in Sweden, California, and elsewhere, while arguing in favor of the original concept. Like most advocates of Human GE, he focuses on parental choice, as does another not afraid to use the word, the bioethicist Arthur Caplan, who wrote, "The most likely way for eugenics to enter into our lives is through the front door as nervous parents—awash in advertising, marketing and hype—struggle to ensure that their little bundle of joy is not left behind in the genetic race.[10]

What Caplan and Galton correctly identify is that state compulsion and market seduction are merely two different means of achieving the same eugenic ends. Modern advocates of Human GE do fit into the tradition of eugenics, whether they like it or not; they are bringing the concept back to its roots.

■ THE GROWTH OF ORGANIZED EUGENICS ■

FRANCIS GALTON, an eccentrically brilliant upper-class Englishman, coined the term "eugenics" in 1883, from the Greek for, roughly, "well bred." He defined it as, "The study of the agencies under social control that make, improve or impair the racial qualities of future generations either physically or mentally."[11] For a long time before he coined the term, Galton had been advocating related ideas. In 1869 he had written *Hereditary Genius: An Inquiry Into Its Laws and Consequences*, in which he proposed arranged marriages between "men of distinction and women of wealth" to produce a gifted race.[12]

He was a man of many and varied interests, but he stuck with this one and, in 1908, founded the Eugenics Education Society of Great Britain. This later became the Eugenics Society, and still exists, having been renamed, in a truly classic piece of euphemism, the Galton Institute. Within six years it had over a thousand members, many of them extremely distinguished, and branches in half a dozen British cities. In 1912, the year after Galton's death, the First International Congress of Eugenics was held in London.

A single libertarian MP, Josiah Wedgwood, was largely responsible for foiling efforts to pass eugenics laws in 1913. Later, the British Labour Party was instrumental in preventing them, but many British progressives of the era were in general strong supporters of eugenics.[13] So were right-wing figures in both Britain and the US and much of the establishment in what were then the two most powerful countries in the world. Eugenics was well on the way to becoming the "conventional wisdom" of the day (see **Box 8.1**).

After the First World War, the focus of activity moved across the Atlantic, where the Second and Third International Congresses were held in New York in 1921 and 1932 (there has been no fourth, yet). It was in the US that eugenics had its most important early legal successes, including on the federal level a particularly terrible Immigration Act (see **Box 8.2**).

By the mid-1920s, eugenics societies were thriving in countries from Belgium to India and from Brazil to Japan. Delegates from a score of countries attended the Second Congress, even though the Germans and Russians were not invited, in the aftermath of war and revolution. The Germans, however, quickly paid attention to the US experience (see **Box 8.3**).

8.1 WHO SUPPORTED EUGENICS?

A REMARKABLE CROSS-SECTION of mainstream intellectuals supported eugenics in the early twentieth century. Among the prominent supporters were:

- Winston **Churchill**, a Vice President of the First International Congress[14]
- Alexander Graham **Bell**, Honorary President of the Second International Congress
- Herbert **Hoover**, who attended the Second International Congress
- Bertrand **Russell**, philosopher and pacifist activist
- J. P. **Morgan**, the world-renowned capitalist
- John Maynard **Keynes**, the world-renowned economist
- Oliver Wendell **Holmes**, a Justice of the Supreme Court
- H. G. **Wells**, novelist, activist and writer of educational non-fiction
- Sigmund **Freud**, who said eugenics was "for the betterment of the world."[15]

That is only a selection of those whose names remain well-known eighty or more years later. To the list can be added, among others:

- the President Emeritus of Harvard
- the President of Stanford
- the Director of the American Museum of Natural History
- much of the British Fabian Society, which was committed to socialism, though not revolution

Many, though not all, of the eugenics advocates also happened to be the beneficiaries of inherited wealth. The Carnegie, Rockefeller, Harriman, Kellogg, DuPont, Gamble, Ford, and Dodge families, and their foundations, were all involved.[16]

These people are no worse than eighteenth-century slave owners like Thomas Jefferson. Their opinions were conventional at the time.

8.2 EUGENICS AND US IMMIGRATION LAW

CRITICISM OF EUGENICS tends to focus on the involvement of governments in discrimination, sterilization, and murder. The Nazi regime was certainly the worst in this respect, but well before they came to power eugenics was a major and entirely conscious influence on the **1924 U.S. Immigration Restriction Act**. This was designed to keep out:

- Jews (83 percent of Jewish immigrants had been classified as "feeble-minded" by an early IQ test)
- Italians (who had a "tendency to personal violence")
- Other "socially inadequate" groups.

The "inadequate" were defined by the **"expert eugenics agent"** appointed by the House Committee on Immigration and Naturalization, Harry Laughlin of the Eugenics Record Office (see also **Box 8.3**).

Calvin Coolidge summed up the principle behind it, as he signed the bill into law,[17] "America must remain American."

That quota system remained in place, with some modifications, until it was replaced by the 1965 Immigration and Nationality Act.

The American branch was led, if we are to single out an individual, by Charles Benedict Davenport, who founded the Eugenics Record Office, which functioned from 1910 to 1940 at the Cold Spring Harbor Laboratory, which has since repudiated without denying its past. Many histories of the movement are available, some listed at the end of this chapter, but one episode will stand as an example—and it is only an example—of what eugenics meant to people here, within living memory.

■ THE CASE OF CARRIE BUCK AND HER FAMILY ■

EUGENICS WAS NOT just a matter of sweeping generalizations. When Oliver Wendell Holmes, Jr., issued his notorious

8.3 NAZI EUGENICS: THE AMERICAN CONNECTION

> It's startling how much Hitler idealized American eugenics.
> —Edwin Black, author of War Against the Weak[18]

THE CONNECTION BETWEEN the German Nazis and the American eugenicists is no secret, and never was. Both sides boasted about it. Nevertheless, the publication of a new book on the subject—such as Black's extensively documented 2003 work—still seems to cause shock, presumably because we do not want to believe it.[19]

The **German sterilization law of 1933** resulted in the sterilization of over 400,000 people.[20] One of its principal sources was the model law developed by Harry Laughlin, the Immigration "expert" from the Eugenics Record Office (see **Box 8.2**). Laughlin originally published his version in 1914, advocating the sterilization of the "feebleminded, insane, criminalistic, epileptic, inebriate, diseased, blind, deaf; deformed; and dependent [including] orphans, ne'er-do-wells, tramps, the homeless and paupers."[21] He lobbied energetically for his concept at the state level, where he was often pushing on an open door, since twelve states already had such laws and more than twenty others—thirty-three in all—were to pass them.

Laughlin was fully aware, at the time, of the Nazi admiration for his work. He had the German law translated into English for publication, and in 1936 accepted an honorary degree from the University of Heidelberg for his work in *"the science of racial cleansing."*

This was by no means a solo effort. The Rockefeller Foundation was involved, as was the Carnegie Institution, and there were reports of the German efforts in, for example, the Journal of the American Medical Association. The Nazis didn't invent eugenics. They just carried it through to one of its logical conclusions.

judgment that, "Three generations of imbeciles are enough," he was not referring to a race or culture or anonymous group. He meant **Emma, Carrie, and Vivian Buck**. The experience of the Buck family—including Carrie's sister **Doris**—is worth examining not because it is unique but because it was all too typical.

Holmes and his colleagues on the US Supreme Court were upholding the Virginia Eugenical Sterilization Act of 1924, which would prevent 8,300 allegedly "defective persons" whose reproduction represented "a menace to society" from having children (or more children if they were already parents), but he was doing so in reference to a particular individual, Carrie Buck, on whose behalf the appeal had been made.

The Court was given the wrong facts on which to decide the constitutionality of the law. At least two and very likely three of the individuals involved were not "imbeciles" at all.[22] From youngest to oldest:

▶ **Vivian Buck**, supposedly diagnosed as feebleminded at the age of seven months, went to school and made the honor roll. Her early school grades were adequate (As in deportment, generally Bs and Cs with the occasional D) until her premature death at the age of eight.

▶ **Carrie Buck**, Vivian's mother, became pregnant at seventeen because she was raped, not because she was "morally delinquent" as claimed in court. She was sent away because of her unmarried pregnancy and therefore supposed "social deviance." She was literate and in her old age enjoyed crossword puzzles.

▶ **Emma Buck**, Carrie's mother, was also an unmarried mother, and may well have been institutionalized for the same reason. Given everything else we now know, it is hard to have confidence in her diagnosis either.

The testimony offered at trial reeked of prejudice, albeit based on class, wealth, and education rather than skin color or familial origin. For example, this is what the superintendent of the institution where Emma Buck resided testified: "These people belong to the shiftless, ignorant, and worthless class of anti-social whites of the South." Harry Laughlin sent a written deposition, although he had never even met Carrie Buck, while a colleague of his, Arthur

Estabrook, examined the seven-month-old Vivian with a Red Cross nurse and pronounced her "below average" and "not quite normal." Estabrook was a sociologist, so why he thought he was qualified to say is unclear; and he was of course wrong.

That Holmes was factually incorrect is shocking and utterly reprehensible but only part of the point. The simple fact is that, for reasons that were supposed to benefit society as a whole, an individual's reproductive rights were removed. There may perhaps be instances where this is appropriate, but the bar to cross is very high indeed; to say that one baby is "not quite normal" does not even come close to being a sufficient justification, even if it were true.

What happened to Carrie Buck's sister is also horrifying. Doris was not part of that lawsuit, and did not even know she should have been. She too was sterilized but she was *never even informed*. (She was told the operation was for appendicitis.) Doris did get married, and she and her husband "wanted children desperately" and always wondered why they could not have them.[23] She only learned the truth in her old age, when researchers began digging up the facts of her sister's case.

Holmes was an individual trying to do his best. Carrie Buck was an individual, too. She might or might not later have chosen to have sex, or to use birth control or even abortion if she did. That was none of the court's business.

One individual was making a judgment about another individual *who did not even exist*—Carrie Buck's second child. Would her unconceived child be "defective"? How about "superior"?

■ POSITIVE EUGENICS ■

THE APPALLINGLY LOGICAL corollary to the Buck decision would be to insist that talented couples do have children together, and the more the merrier. That is exactly what Francis Galton wanted. He was by inclination a "positive" eugenicist who hoped to increase the frequency of "superior" genes, which in his view meant those found in the British upper classes, the subject he had investigated in *Hereditary Genius*.

The difficulty, of course, was enforcement. In fact, the prevalence of "negative" eugenics—sterilization and outright murder—arose largely because no one could figure out an adequate method to implement the positive version. That is exactly the problem that the modern advocates of Human GE hope to be on the verge of solving. They don't need to make sure that egg and sperm both come from "superior" beings—they will just "fix" the end product.

If they claim that their project differs in principle from the old eugenics, they are precisely wrong: They are, in fact, using modern technology to revive Galton's original concept.

■ NEGATIVE EUGENICS ■

NEGATIVE EUGENICS has been thriving, on an individual level, though many people probably do not realize that they are practicing it. All abortions for reasons affecting the predicted health of the embryo or the person into which it might develop are eugenic by definition. So too are all forms of embryo selection, which is routinely practiced during IVF treatment (See **Chapter 2**).

There are those who oppose all abortions and all embryonic selection on principle; for them the line as to what is permissible is simple to define. For most of us, however, it is not; and it must be stressed that *eugenics is not the kind of proposition that is proved or disproved by a single example. It is a mind-set and a way of approaching the human experience, and as such its boundaries are inevitably gray.*

Some advocates of modern genetic engineering seem to believe that pointing out that, as a society, we do in fact routinely practice some form of eugenics is sufficient to justify their every proposed abuse—which is nonsense. It is also the kind of palpable nonsense that the public as a whole has no trouble seeing for what it is (see **Chapter 9** on public opinion).

■ BLURRING THE LINES ■

THE BOUNDARY BETWEEN obviously acceptable and distinctly dubious eugenic practices is rapidly becoming fuzzier, with the development and increased use of preimplantation genetic diagnosis (PGD; see **Chapter 2** and **Chapter 5**). Consider these three real-life scenarios:

1. In 1998, PGD was used for the first time *to screen against sickle-cell anemia*.[24] The resulting twins were reported to be healthy and free of the sickle-cell trait, which is presumably a good thing for them. It remains true, however, that the disease is not fully understood, that in some cases the symptoms are not severe, that possession of a single copy of the gene offers some protection against malaria, and that it no longer seems to be a clear-cut case of a single-gene disease.[25]

Was that an ethically acceptable choice? And does anyone have the right to interfere except the parents and their physicians?

2. Shortly afterward, PGD was used *to select an embryo whose tissue would be compatible with a sibling* who suffered from Fanconi's anemia and would benefit from a transplant.[26] On the face of it, the child created from the chosen embryo was born in order to become a donor, long before he was in any position to give (or withhold) consent.

Was that an ethically acceptable choice?

3. A year or so later, PGD was used *to select an embryo that
   did not have a mutation in a specific gene that is related
   to an increased risk for cancer.*[27] The child is still thought
   to have a 40 percent chance of developing cancer, as do
   we all, and an embryo with the mutation is thought to
   have a 10–20 percent chance of living a cancer-free life.
   Even the doctor who performed the test, who believes
   it should only be used to avoid illnesses, "conceded
   that the line can be fuzzy when, as in the case of Li-
   Fraumeni, disease is not 100 percent certain to occur."

Was that an ethically acceptable choice?

The next step is to use the test for something that *never*
constitutes a disease in the normal sense of the word. And
indeed it has already happened—sex selection, as discussed in
Chapter 5. What is next? In the unregulated world of the fertility specialists, almost anything could happen. It sometimes
seems that the virtually total lack of regulation in the US has
left both legislators and the public in a kind of dream state
where onrushing disaster seems literally inevitable. It is not. But
we had better act soon.

■ AFFECTING THE GERMLINE ■

A FEW EXPENSIVE, individual interventions via PGD may not
affect the human germline in a material way—that is, change
the common genetic pool that constitutes humanity as we
know it—but it should be noted that we are already significantly
altering it in both eugenic and dysgenic ways. The increase in
diabetes is an example of a dysgenic one: "Type 2 diabetes is
predicted to increase by 42 percent in developed countries and
as much as 62 percent in the rest of the world by 2025, when
it is projected to affect 300 million people."[28]

8.4   PROGRESSIVES AND EUGENICS

EUGENICS WAS NOT just a conspiracy of the privileged. Certainly, the taxpayers of Virginia and elsewhere didn't want to waste their money on the undeserving poor—but many progressives supported the project.

Bertrand Russell's inclusion in the list of eugenics supporters is particularly notable. A world-renowned logician, and winner of the Nobel Prize for literature, he was also a radical political activist of the first rank. He was jailed for criticizing the First World War and again forty years later, at the age of eighty-eight, in the service of nuclear disarmament. For a contemporary comparison, think perhaps of Noam Chomsky, without tenure but with aristocratic connections and a willingness to put his body on the line.

Russell was a more subtle and less prejudiced advocate for eugenics than most but he proves that eugenicists were not all unspeakable Neanderthals. They were wrong, but often well-intentioned and sometimes both brilliant and radical. That the great logician got fooled by that awful proposition is a salutary lesson for anyone who thinks they are smart enough to make irreversible decisions that will reverberate down our family tree.

Modern liberals are just as liable to fall prey to the delusion. For example, **Ronald Dworkin** explicitly claims that principles of equality and autonomy justify deciding "which kind of people, produced in which way, there are to be."[29] This strange, self-contradictory posture can be summed up as: "We refuse on principle to tell you what to do but we insist that you make the correct decision."

That amounts to a demand that parents intervene before birth to ensure that only "high quality" children are born. **Dr. Robert Edwards**, the man responsible for the first successful in-vitro fertilization (IVF), said in 1999, "Soon it will be a sin of parents to have a child that carries the heavy burden of genetic disease. We are entering a world where we have to consider the quality of our children."[30] Doubtless the word "sin" makes some liberals nervous, because of its overt religious connotations, but he is speaking to the heart of the matter: Eugenics is more than a question of laws, it is at root a question of faith, a belief in science and progress and human control; a faith that those who doubt call hubris.

8.4   Each of the supposedly logical steps that lead to eugenics can and must be challenged. The genetically disabled may not "want" to be disabled but a remarkable number of them argue strenuously against interventions that would have prevented them from existing (see **Chapter 11**). The scientific connections between specific genes and disabilities are rarely definite; it may be that most traits that people fantasize about changing cannot be altered without unacceptable risk (see **Chapter 2**).

Besides, do we know how to perfect our own lives? Can we possibly believe that we know how to improve our entire species?

It is a commonplace that modern medicine allows many individuals to live and reproduce who once would have died in early childhood; in fact, this was one of the motivating forces behind eugenics in the first place. Simplistic eugenicists say we must be becoming weaker as a species; the immediate retort is that many talented individuals are now able to contribute to society when previously they would have perished. These processes will go on, and argument about them is fruitless and, in the end, irrelevant. That we do make *some* changes in the germline is no reason at all for us to allow *every* proposed change.

Positive eugenics is potentially a much bigger concern than negative eugenics. This was less true a century ago, when the worst that could happen was some kind of social influence over mating decisions, than it is now, with the threat of artificial chromosomes and the introduction of non-human genes (see **Chapter 2**). But the idea was there then, and will undoubtedly persist. It has always had, and will continue to have, particular appeal for interventionist progressives (see **Box 8.4**). If history teaches us nothing else, it should be to think twice.

As with negative eugenics, there is inevitably a gray area. Juliet Tizzard, then director of the British Progress Educational Trust, attacked the whole idea of applying the concept of eugenics to modern trends in biotechnology thus: "I personally like tall, dark-haired men, have chosen a partner accordingly and hope my children will reflect that preference. In so doing, I am seeking a modicum of control over my future children. Does that make me a eugenicist?"[31] Well, in a sense, yes. Does it make her a bad person? Of course not. To compare such an ordinary human reaction with Nazi atrocities would be like equating the politeness of minor social hypocrisy with Holocaust denial—the issues are not in the same realm of meaning.

Certainly in some societies, marriages are or have been arranged, partly for reasons of dynastic power, and partly with the thought in mind of the children to be expected. This undoubtedly can represent a eugenic impulse, though not in practice a very powerful one. Francis Galton himself tried to propose a register of outstanding human stock, but his distinguished cousin Charles Darwin, who was skeptical of eugenics (though his son Leonard directed the First International Congress) suggested that "within the same large family, only a few of the children would deserve to be on the register."[32]

There also have been various discussions about encouraging births with tax breaks and government grants, but the eugenic intent founders on the difficulty of deciding who exactly is qualified for them. Probably the most specific proposal was made in Singapore, which once toyed with the idea of restricting "baby bonuses" to university graduates, but dropped the idea as too controversial.[33] China, which enforces severe restrictions on reproduction, insists, to some skepticism elsewhere, that it is opposed to eugenics, "which can easily cause inequality among human beings."[34]

■ MARKET FORCES COMPARED WITH STATE INTERVENTION ■

IT IS INCREASINGLY clear that the principal modern threat of eugenics lies in the "market forces" that, unless checked, may allow those who can afford to do so to experiment on their—and, in the broader sense, our—children. The confusion here is between eugenics itself and state coercion.

What modern eugenics really implies, and most of its advocates are quite clear thinking enough to agree, is that instead of the poor being prevented from breeding, which is what in practice the old eugenics tried to do, the rich would be encouraged to have "designer babies"—positive eugenics.

One, only somewhat metaphorical, way of illuminating the contradiction is to say that, to these advocates, market forces effectively *are* the state. They completely fail to see that markets are not natural forces like gravity but social phenomena, which are manipulated and regulated, both by governments and by large participants in them. In transferring the mechanism of eugenics from the state to the market, nothing fundamental is altered.

People haven't changed since the days of Homer, and probably long before. Cassandra, the prophet who was never believed, was a princess of Troy, and the main lesson to be learned from history may be that we don't learn lessons from history. But it is important that we place the modern eugenicists squarely in the tradition where they belong.

Of *course* they think they are doing good; that is what eugenicists have always thought. Certainly they are intelligent and often distinguished people; so were most of those who developed the idea in the first place. Undoubtedly they see their ideas as inevitable; enthusiasts usually do. Naturally they dissociate themselves from the obvious evils of the past; who wouldn't? But if Francis Galton could have attended the Symposium on Engineering the Human Germline held at UCLA in March 1998, there is no question that he would have felt right at home; these were the people carrying forward his great idea.

■ FURTHER READING ■

Free Documents from the Web

The former home of the Eugenics Record Office, the Cold Spring Harbor Laboratory, now in penance hosts an excellent archive including critical analysis and a substantial bibliography, at www.eugenicsarchive.org/eugenics. Articles include: Garland E. Allen, "Flaws in Eugenics Research"; Paul Lombardo, "Eugenics Laws Restricting Immigration" and "Eugenic Sterilization Laws"; and Jan Witkowski, "Traits Studied by Eugenicists."

A more provocative exploration of the history and issues from a radical Christian viewpoint, with a particular focus on current "crypto-eugenics," is at www.eugenics-watch.com.

Barry Mehler, "Brief History of European and American Eugenics Movements," excerpted from his Ph.D. thesis for the University of Illinois and posted at www.ferris.edu/isar/arcade/eugenics/movement.htm.

Ruth Hubbard, "Eugenics, Reproductive Technologies and 'Choice,'" *Genewatch*, 01/01, brings the story up to date, at www.gene-watch.org/gene watch/articles/14-1choice.html.

Alexander Cockburn and Jeffrey St. Clair, "Eugenics: The Impulse Never Dies," *CounterPunch*, 03/06/00, www.counterpunch.org/eugenics.html

Books

Edwin Black, *War Against the Weak: Eugenics and America's Campaign to Create a Master Race*, New York: Four Walls Eight Windows, 2003

Daniel Kevles, *In the Name of Eugenics: Genetics and the Uses of Human Heredity*, Harvard University Press, 1985, revised 1995, is the classic history.

Stephen J. Gould, *The Mismeasure of Man*, W.W. Norton, revised 1996, is not strictly about eugenics, though it touches on the subject, but assaults the eugenic mindset with unparalleled rigor.

David Galton, *In Our Own Image: Eugenics and the Genetic Modification of People*, London: Little, Brown and Company, 2001, is an attempt to update and rehabilitate the concept by a British Professor of Genetics who appears to be a liberal.

Richard Lynn, *Eugenics: A Reassessment*, Praeger Publishers, 2001, is included (unread) mostly to prove that it exists. A former professor of psychology at the University of Ulster, he's in favor not of genocide but of "phasing out" what he has called "the population of incompetent cultures."

■ ENDNOTES ■

1 Dave Reynolds, "Easley Is Third Governor To Apologize For Sterilizations," *Inclusion Daily Express*, 12/16/02

2 Alexander Cockburn and Jeffrey St. Clair, "Eugenics: The Impulse Never Dies," www.counterpunch.org, 03/06/00

3 ibid.

4 See www.vector.cshl.org/html/eugenics, and James D. Watson, *A Passion for DNA: Genes, Genomes, and Society*, New York: Cold Spring Harbor Laboratory Press, 2000

5 Richard Lynn, *Eugenics: A Reassessment*, Praeger Publishers, 2001

6 Richard Lynn and Tatu Vanhanen, *IQ and the Wealth of Nations*, Praeger Publishers, 2002

7 David Duke, *My Awakening: A Path to Racial Understanding*, Free Speech Books, 1998

8 First International Conference on "Ethics, Science and Moral Philosophy of Assisted Human Reproduction," 09/30–10/01/04; program at www.humanreproethics.org/scintific.htm (typo in the original URL)

9 David Galton, *In Our Own Image: Eugenics and the Genetic Modification of People*, Little, Brown and Company, London, 2001

10 *Time*, 01/21/01

11 Francis Galton, *Inquiries into Human Faculty and its Development*, J. M. Dent and Sons, London, 1883

12 Francis Galton, *Hereditary Genius: An Inquiry Into Its Laws and Consequences*, St. Martin's Press, NY, 1978; first published 1869

13 David Galton, op. cit., p. 90; *National Review*, 07/37/00; John Macnicol, "Eugenics and the Campaign for Voluntary Sterilization in Britain Between the Wars," *Social History of Medicine*, 08/89, pp.147–169

14 Churchill, whose career was long and not without its contradictions, seems to have changed his mind by 1940, when he referred to "the abyss of a new dark age made more sinister, and perhaps more prolonged, by the lights of a perverted science" (quoted in *The Scotsman*, July 4, 2000). Faced with the Nazis, many did. The other references are from sources cited above, and especially www.eugenics-watch.com.

15 Freud was quoted in the *New York Times*, 02/16/01.

16 See Barry Mehler, "Brief History of European and American Eugenics Movements," PhD Thesis for the University of Illinois.

17 Paul Lombardo,"Eugenics Laws Restricting Immigration," www. vector.cshl.org

18 Quoted in *USA Today*, 09/14/03

19 Edwin Black, *War Against the Weak: Eugenics and America's Campaign to Create a Master Race*, Four Walls Eight Windows, New York, 2003

20 *Science*, October 5, 2001

21 Paul Lombardo, "Eugenic Sterilization Laws" at www.vector.cshl.org

22 Stephen Jay Gould, *The Flamingo's Smile: Reflections in Natural History*, W. W. Norton & Company, 1985

23 Quoted in David Galton, op. cit.

24 *Journal of the American Medical Association*, 05/12/99

25 Francis S. Collins, Lowell Weiss, and Kathy Hudson, "Have No Fear. Genes Aren't Everything," *The New Republic*, June 25, 2001

26 London *Independent*, 10/05/00

27 *Washington Post*, 06/09/00

28 *EurekAlert*, 09/28/01

29 Adam Wolfson, "Politics in a Brave New World," *Public Interest*, Winter 2001.

30 Quoted in *Business Week*, 07/15/99

31 *BioNews*, 10/18/99

32 David Galton, op. cit., pp. 83–5

33 Singapore's first and longtime Prime Minister, Lee Kuan Yew, contributed an article titled "Differential Fertility and Population Quality" to *Population and Development Review* in 1983, and argued that high birth rates among the "intelligent" should be encourged.

34 Xinhua News Agency, 07/08/00

9

MANIPULATING PUBLIC OPINION

■ INTRODUCTION ■

P EOPLE DON'T LIKE biotechnology. They are willing to pick and choose some possible benefits, but, according to the National Science Foundation (NSF) annual survey for 2004,[1] "In no NSF survey year has a majority of Americans agreed that the benefits of genetic engineering outweigh the harmful results." Similarly the third annual Virginia Commonwealth University (VCU) Life Sciences Survey was headlined, with delicate understatement:[2] "Public Values Science But Concerned About Biotechnology."

All four VCU surveys conducted from 2001 to 2004 showed majority agreement with the statement "Scientific research has created as many problems for society as it has solutions" (51–45 in 2004, down from a high of 59–39 in 2002). Genetic testing is generally favored, although there is concern about what is seen as likely discrimination. So are some applications, and the continuance of genetic research, but overall: "NSF survey data show a slight, gradual decline in the American public's support for genetic engineering between 1985 and 2001. The shift can be seen most clearly among college-educated respondents and those classified as attentive to S&T [science & technology] issues." That last sentence may the most significant of all. If genetic engineering is losing its allure among those most likely to support it—and also most likely to vote and influence policy—then we may be able to get appropriate regulations to curb abuses.

■ FIXED AND FLUCTUATING VIEWS ■

BEFORE DRAWING CONCLUSIONS from *any* poll, it is absolutely vital to read the questions with great care—and that goes double for summaries released to the press, and triple for headlines based on the handouts. This is especially true when the poll was commissioned by a campaigning organization (see **Box 9.1**), and even more so when the results were not what the sponsor hoped for (see **Box 9.2**).

9.1 SLANTING THE POLLS

IN THE EARLY SUMMER of 2001, the issue of stem cells was new to the wider political audience. An important topic was whether the federal government should fund research on embryonic stem cells. Partisans on both sides of the issue conducted opinion polls, with wildly different results:

■ One found **70 percent opposition** to federal funding
■ One found **70 percent support** for federal funding

These two polls, both scientifically adequate samples, were conducted within three weeks of each other, in May and June. To the surprise of no one, the results confirmed the opinions of the people who commissioned them. How did they do this?

■ The one conducted for the National Conference of Catholic Bishops told participants that "live embryos would be destroyed in their first week of development" and asked about "using your federal tax dollars for such experiments."[3]
■ The other, conducted for the Coalition for Advancement of Medical Research (CAMR), avoided the term embryos completely, but referred to "excess fertilized eggs," and listed seven "deadly diseases" the research could help treat.[4]

SPINNING THE RESULTS

IN 1999, Novartis commissioned an investigation into British public attitudes toward the life sciences, conducted by the reputable pollster MORI. Even an industry organization couldn't avoid the conclusion that,[5] "The poll found widespread opposition to many new technologies." But they did make a rather desperate attempt at positive spin: "However where the benefits were clearly communicated there was generally an increase in public support." A closer look shows that what was meant by "clear communication" was attaching the qualifying phrase, "If a permanent cure or vaccine for Alzheimer's disease could be found" and asking the question again.

But the public still would not support animal cloning or the genetic modification of either plants or animals. The best the pollsters could get, as described in their published report, was, "Approval ratings for genetic modification of plants increased if it was proved to be necessary . . . to obtain nutritionally improved food that tastes and costs the same as current food (from 20 percent to 34 percent)."[6] Hmmm, better food for the same price and people still didn't want GE? Perhaps there was a problem with the public relations campaign.

It is clear that public opinion on some biotechnology topics is well-formed and fixed. For example, by overwhelming and consistent majorities—about ten to one—people are *against* reproductive human cloning (see below) and *in favor of* genetically modified food being labeled.[7] Strangely, however, cloning is not illegal in most of the US, and labeling is not required.

A smaller, but stable, majority is adamantly opposed to animal cloning—opposed enough to think it should be illegal. Research continues, however, although the success rates are terrible (see **Box 3.3**). In fact, cloned animals are beginning to be developed commercially with little opposition in mainstream political circles.

On other topics, public opinion seems to be evolving and may be heading toward a politically very difficult position, in which a significant minority remains deeply offended by what the majority considers appropriate or even desirable. That is, of course, roughly a description of the situation with regard to abortion, and it may be how part of the embryonic stem cell debate (see **Chapter 4**) will eventually play out. The best guess, discussed below, is that opinion is not yet settled.

On the broader, longer-range issues of germline intervention and enhancement, it seems even more likely that public opinion is not fully formed. Certainly the advocates of Human GE hope so, because all the indications are that people *reject* the idea of "improving" humans. Picking eye or hair color—the choices pollsters tend to use, presumably because they are easy to understand—is definitely unpopular. Some kinds of disease-avoidance might, however, be accepted; the questions are often loaded and sometimes vague.

The available polling data does indicate that public opinion on biotechnology varies significantly by religion, education, gender, race, and income. To be blunt: rich, white men with graduate degrees for whom religion is not very important are far more likely to approve of cloning than are poor, black women who didn't go to college but regularly go to church. That's on average, and with exceptions, of course; most of us fall somewhere between these stereotypes. The political consequences of this are extremely important, and briefly discussed at the end of the chapter.

▪ MEDICALIZING THE BUSINESS OF RESEARCH ▪

THE GENERAL MOOD of skepticism about biotechnology is undeniable. Its advocates therefore do their best to accentuate the positive, by associating their product with things that people do like, and polling to see what works before rolling out a public relations campaign. Thus, feeding the starving is generally seen as a worthy goal, so GE food is regularly (and quite falsely) claimed to be the way to do it.[8] And what is the equivalent for Human GE? Healing the sick.

Anything that promises us a new medical technique is bound to be viewed favorably. Hence the references to "therapeutic" cloning and emphasis on "cures" in every discussion about research priorities. Hence, too, language like "medical research" in polls discussing attitudes to what is really basic science.

We still trust doctors. For example, a Gallup Poll in late 2003 gave doctors a 68 percent "high or very high" rating on honesty and ethical standards, behind only nurses (who got a remarkable 83 percent) and well ahead of the clergy (56 percent), let alone bankers, journalists, politicians, HMO managers (9 percent) and car salesmen (7 percent).[9]

Obviously, this is not lost on genetic researchers, who seem to have successfully aligned themselves with doctors, at least for the moment. The confusion of the PhD with the medical degree certainly helps, but it remains to be seen how long this will last, if "gene therapy" and the like (see **Chapter 6**) keep over-promising and underdelivering.

There are also many examples of what are either actual conflicts of interest (see **Box 6.2**) or potential conflicts of interest. Often, no one needs to lie, or even dissemble; they simply don't mention them, and journalists rarely call them on it. Neil Munro, one of the few who have, cites the example of Doug Melton of Harvard. Melton is a genuine expert on embryonic stem cells, who was quoted in thirty-six newspaper or magazine articles in less than two years. Only *one* (written by Munro himself) mentioned that Melton was *also* the chief scientist at Curis, Inc., which he cofounded and which stands to profit from that technology.[10]

Knowing such things might bias the public's opinion, of course, and we wouldn't want to do that.

■ TRUST US, WE'RE EXPERTS ■

A VERY COMMON attitude among scientists (and, to be fair, most people) is that those who disagree with their judgments lack the knowledge required to make the correct choice. It is not surprising, therefore, that several surveys have concluded, perhaps a little hastily, that the public doesn't know enough about biotechnology.

Enthusiasts for GE take comfort from that—they often assert that with familiarity and education will come acceptance. For example, they say, IVF was unpopular at first, but people came round to accepting the technology as they got used to it. This is just not true (see **Box 9.3**). On the contrary, the public, collectively, seems quite able to make distinctions between acceptable and unacceptable technologies.

9.3 | **THE (LACK OF) EVOLUTION OF PUBLIC OPINION ABOUT IVF**

FERTILITY TREATMENT IS widely, though not unanimously, supported. According to a 2002 poll conducted for the Genetics and Public Policy Center, 72 percent of Americans (74 percent of women, 70 percent of men) approve of in vitro fertilization (IVF) and only 20 percent disapprove.[11] The same poll revealed that almost a third (29 percent) had either tried IVF themselves or knew well someone who had. Clearly it's now part of our society's common landscape.

Human GE proponents like to claim that procedures we now consider beyond the pale will also become accepted. Rael, the prophet of extraterrestrial cloning (see **Box 3.6**) said just that to Congress: "It was exactly the same problem . . . when IVF started."[12] Robin Marantz Henig wrote that "our collective attitude about in-vitro fertilization took a 180-degree turn after the first few test-tube babies."[13] That's really not what happened.

The first "test tube baby" was born on July 25, 1978, to a worldwide blaze of publicity. The lead headline in the New York Times was typical: "Scientists Praise British Birth as Triumph." A few theologians did object; some others fretted about safety and the allocation of scarce medical resources; the farsighted (and progressive) British MP Leo Abse worried that "we are moving to a time when an embryo purchaser could select

9-3

in advance the color of the baby's eyes and its probable IQ." But IVF itself was accepted almost at once.

The public was fascinated—by November, 93 percent were aware of the birth, and the polls show that people definitely approved:

- A September 1978 Harris poll said that 85 percent of American women over 18 years old thought the "test-tube" method should be available to married couples that are unable to have children.[14]
- A November Gallup poll found 60 percent in favor of IVF, 27 percent opposed, and 13 percent with no opinion.[15]

In other words, society collectively made up its mind quickly and decisively, and hasn't changed much in a generation.

But in the face of public disquiet, there are regular reports that surveys indicate that the public doesn't know what it's talking about. Two of them are examined in **Box 9.4** and **Box 9.5**, and both seem less conclusive than their summaries (and press releases) suggest. Moreover, researchers trying to understand public mistrust of biotechnology discovered,[16] "contrary to common belief, that being better informed about biotechnology is not a precursor to being more open and accepting of it."

9-4

AN INDUSTRY POLL ON PUBLIC AWARENESS

THE PRESENTATION based on a poll conducted in 2004 for the Biotechnology Industry Organization (BIO) stressed that:[17]

- One in five consumers admit they have never heard the word "biotechnology."
- Consistently, fewer than half of consumers say they've heard any more than a little about biotechnology.

The survey was intended to highlight the differences in opinions between "working biologists" and "consumers." This was one "knowledge" question, with responses:

9·4

How informed would you say you are about the intended purposes of stem cell research—very informed, somewhat informed, not too informed or not at all informed?

	CONSUMERS	SCIENTISTS
Very informed	9%	46%
Somewhat informed	53%	44%
Not very informed	26%	8%
Not at all informed	11%	2%

There is an apparent contradiction here. If only 44 percent of the public have heard and read "a lot or some" about biotechnology, then how come 62 percent are very or somewhat informed about the intended purposes of stem cell research? Another conclusion would be that nearly two-thirds of the public has a pretty good idea of what's happening.

9·5

PASS OR FAIL—WHO'S GRADING THE TEST?

AN EXTENSIVE POLL conducted in 2002 for the Genetics and Public Policy Center showed, according to the Center's summary, that,[18] "The public's knowledge about these technologies is not keeping pace with the steep growth in genetic science. Only 18 percent of respondents were able to correctly answer 6 or more of the 8 knowledge questions."

That sounds bad; but another way of reporting the result is to say that 61 percent answered at least half of the questions correctly. And for "6 of the 8 knowledge questions," if you don't count abstentions, at least a plurality got the answer right!

Is that a pass or fail for the American public? (We report, you decide.)

■ HUMAN REPRODUCTIVE CLONING POLLS ■

PUBLIC OPINION ON reproductive cloning—cloning to make babies—is quite remarkably consistent, especially on the basic principle of human cloning.

The wording in the first group of polls that follows varies only slightly. Some ask whether cloning is "morally acceptable" or "morally wrong," others if "you approve or disapprove" of cloning, or "favor or oppose" it.

Poll	Date	Approve (%)	Disapprove (%)	Other (%)
VCU	09/04	13	83	4
Gallup	05/04	9	88	3
VoM	03/04	9	87	3
VCU	09/03	13	84	2
Gallup	05/03	8	90	2
VCU	09/02	16	81	3
Gallup	05/02	7	90	3
Gallup	05/02	8	90	2
Fox	02/02	7	89	4
Gallup	11/01	9	88	3
VCU	09/01	14	82	3
Gallup	05/01	9	89	2
Fox	04/01	6	90	4
Time/CNN	02/01	7	90	3
Gallup	02/97	10	87	3

VCU is the Virginia Commonwealth University Life Science Survey, which asked about cloning after a sequence of questions about medical research; perhaps that accounts for the consistent, though slight, difference—but the ratio of opponents to supporters is still five or six to one. (For Americans to agree by five to one on *anything* is remarkable, and that's the lowest ratio in the list; the highest is fifteen to one, most of them around ten.) VoM is the "Voice of Mom," a survey of adult women. Both VCU and VoM also measured intensity of opinion, registering 4 percent and 2 percent "strongly" in favor, 65 percent and 75 percent "strongly opposed," respectively.

When the questions are specifically about regulation and whether scientific research should be restricted, some minor differences do arise (the answers again follow the pattern *support or allow cloning / oppose / other*):

POLL	DATE	SUPPORT OR ALLOW CLONING (%)	OPPOSE (%)	OTHER (%)
Should cloning be legal or illegal?				
Gallup	01/03	11	86	3
Should scientists be allowed to try to clone human beings?				
CBS	05/02	11	85	4
Do you favor or oppose scientific experimentation on the cloning of human beings?				
Pew	02/02	17	77	6
Should the government have regulations to limit the cloning of humans?				
GPPC	10/02	11	84	5
Do you approve or disapprove of scientists working on ways to clone humans?				
GPPC	10/02	18	76	5
Should we allow "unrestricted scientific research related to human cloning"?				
Pew	03/01	13	81	6

Legislation—specifically an outright ban—does, however, seem to be viewed with some concern by the public. The next two seem at first blush to contradict what we have just seen:

POLL	DATE	OPPOSE	FAVOR	NOT SURE
Would you favor or oppose a law that would prohibit the cloning of human beings, or are you unsure?				
Gallup	04/02	36	39	26
"Do you favor or oppose an outright ban on the cloning of human beings?"				
NBC/WSJ	01/02	39	54	7

The resolution of this apparent contradiction seems to be that confusion has been caused by the relentless promotion of "therapeutic cloning" or "research cloning" (which in turn may have been the intention of some of the people pushing those terms).

Consider the following unusually detailed question posed in January 2003, by the *Los Angeles Times*:

Which of these statements comes closest to your view on human cloning? I support a complete ban on all research into human cloning without exception. I support a ban on human cloning that would still allow research on cloned embryos to learn more about diseases. I oppose any law that restricts research into human cloning.

Complete Ban	43
Partial	41
Oppose Restrictions	11
Don't Know	5

That still adds up to at least an 84 to 11 (almost 8 to 1) overwhelming majority for banning reproductive cloning.

■ POLLING ON EMBRYONIC STEM CELL RESEARCH ■

PUBLIC OPINION ON embryonic stem cell research, in contrast, may not be settled. Setting aside the transparently biased polls (see **Box 9.1**), there is still a wide range of responses, which may reflect more subtle, even unintentional, biases. All of these were conducted by reputable organizations in the year following the campaigners' ones, and asked about federal funding of stem cell research:[19]

POLL	DATE	FAVOR	OPPOSED	DON'T KNOW
Beliefnet	06/01	60	31	9
Gallup	07/01	30	13	57
Harris	07/01	43	27	30
ABC	07/01	60	36	3
Gallup	08/01	55	29	16
Pew	02/02	43	35	22

The Beliefnet one focused on "medical research" and so did the first Gallup, but the latter specifically added "or don't you know enough to say?"—and respondents seem to have told the truth! Harris asked for an opinion on funding "based on what you have read or heard," while the ABC was essentially a repeat of the Beliefnet (they cosponsored both) and got similar results. Pew asked simply, "Do you think the federal government should or should not fund stem cell research?"

The second Gallup differed from the first in that it left out the "don't know" clause, which made a huge difference. It did, however, proceed to ask about funding research on adult stem cells (supported 68–26), cloned embryos (opposed 28–65), and embryos created with standard IVF techniques but specifically for research purposes (opposed, but only 46–49).

Naturally, it also helps to mention a particular disease, as can be seen from these two 2004 surveys, taken less than a month apart, the second of which included a crude breakdown by political affiliation:

Group	Favor	Opposed	Don't Know
"Do you support or oppose federal funding for embryonic stem cell research?"[20]			
All	43	47	10
"Do you favor or oppose federal funding of research on diseases like Alzheimer's using stem cells taken from human embryos?"[21]			
All	64	28	8
Republicans	53	38	9
Democrats	74	20	6
Independents	67	26	7

Moving away from the issue of federal funding, the VCU survey has consistent methodology and questions, but somewhat fluctuating results, as do others:

POLL	DATE	STRONGLY FAVOR	SOMEWHAT FAVOR	SOMEWHAT OPPOSE	STRONGLY OPPOSE	DON'T KNOW
VCU	09/04	24	29	14	22	11
VCU	09/03	17	30	21	23	9
VCU	09/02	12	23	22	29	15
VCU	09/01	17	31	21	22	9

On the whole, how much do you favor or oppose medical research that uses stem cells from human embryos . . . ?

Dr. Matthew Nisbet, of Ohio State, has documented the evolution of polling on stem cells (and also reproductive cloning), in great detail. With academic precision and delicate understatement, he concluded in 2004 that,[22] "question wording in surveys can have strong effects on the public's stated response to these volatile issues." As to the issue of stem cells in particular: "a substantial proportion of Americans remain unsure about the matter."

Perhaps that first Gallup poll is the one that got it most right: 57 percent don't know.

■ CHOOSING OUR CHILDREN'S TRAITS? ■

This is the key issue—the single most important question in the Human GE debate. Do we initiate changes in the germline (see **Chapter 2**)?

The standard media catchphrase is "designer babies," but that has been overused, for example to describe children chosen after the use of pre-implantation genetic diagnosis (PGD) in order to be possible tissue donors, usually for siblings. They are *selected* children, but not really *designed*. That practice is certainly controversial; one poll suggests the public supports it by 61–33, but the exact wording is not available.[23]

As ever, wording makes a big difference. For example, the 2002 VCU survey tried to use one catchall question to cover this subject:

New technology in science and medicine may allow couples who want to have a baby to pick and choose the baby's genetic characteristics such as hair color or the risk for certain diseases. Do you strongly favor, somewhat favor, somewhat oppose, or strongly oppose picking and choosing an unborn baby's characteristics using these new technologies?

Strongly Favor	Somewhat Favor	Somewhat Oppose	Strongly Oppose	Don't Know
5	13	21	58	4

In 2003, however, the same survey asked two different questions, and although they did not break down the results by intensity of feeling, the answers are revealing:

Would you say that changing a baby's genetic characteristics for cosmetic purposes such as eye or hair color is making appropriate use of medical advances OR is it taking medical advances too far?

Appropriate	Too Far	Don't Know
4	94	2

This response was completely consistent across social and educational categories—no group gave less than 92 percent opposition. The highest levels came from those who think abortion should always be illegal (98 percent) and those with at least an undergraduate degree (97 percent), two groups whose views do not usually overlap this closely. Religious views made no difference whatsoever—actually, those who do not consider religious guidance very important registered fractionally more opposition than the strongly devout, but only 95–94.

Would you say that changing a baby's genetic characteristics to reduce the risk of serious diseases is making appropriate use of medical advances OR is it taking medical advances too far?

Appropriate	Too Far	Don't Know
41	54	6

Note that, once again, an emphasis on "medical" applications—here, "advances," which is even more loaded, used twice for an even greater effect—makes a large difference in the results. They did reveal some variations, with the highly educated and the irreligious regarding such interventions as *appropriate* (53–42 and 52–43), while the devout *opposed* them (33–59) and those who believe abortion should always be illegal were most strongly opposed (21–76). (For more on this, see the discussion of opinions on animal cloning, below.)

These responses seem to be fairly typical:

- A 2003 Gallup survey came out 89–8 against allowing parents to select "genetic traits such as intelligence, height, or artistic talent" and an even greater 91–7 against selecting the traits for "your child."[24]
- The 2002 GPPC survey showed 72–22 opposition even to *testing* during pregnancy to make sure your baby "has desirable characteristics such as high intelligence and strength" (which is not possible anyway).
- Testing for "a tendency to develop a disease like cancer when he or she is an adult," however, was approved 60–33.
- The same poll came out 76–20 against parents being "offered a way to change their *own* genes in order to have children who would be smarter, stronger, or better looking" but 59–34 in favor of doing so "to prevent their children from having a genetic disease."

Some advocates of Human GE, including Gregory Stock (see **Chapter 10**), cite 1993 polls to suggest that people have a much greater willingness to use genetic engineering "both to prevent disease and to improve the physical and mental capacities inherited by their children. The numbers ranged from 22 percent in Israel and 43 percent in the United States to 63 percent in India and 83 prcent in Thailand."[25] The trouble is that these were conducted before gene therapy had failed (see **Chapter 6**); they were devised and reported under the title "Public Acceptance of Human Gene Therapy and Perceptions of Human Genetic Manipulation."[26] In other words, the assumptions on which the questions were based are far out of date (they were already almost a decade old when Stock cited them), and the statistics are therefore essentially rigged.

There are important issues of principle involved in the decision about where to draw the line in Human GE. The *tactic* of selling one position under the guise of "medical advances" is one that the public should keep in mind. It will definitely be used again and again.

■ MISCELLANEOUS POLL DATA ■

The following is a small, and necessarily incomplete, selection of poll results that seem to be at least indicative of public opinion on various related subjects. For more specific information, follow the web links listed at the end of the chapter.

The first, from the 2003 VCU survey, might perhaps seem surprising, especially to the **longevity** lobby:

New genetic techniques may prove able to slow down the aging process in human beings. How likely would you be to use genetic therapies if it meant you could live longer?

VERY LIKELY	SOMEWHAT	NOT TOO	NOT AT ALL	DON'T KNOW
14%	23%	25%	36%	1%

Combined, for simplicity, the result is then 37–61 *against*. This was fairly consistently true among different groups; the biggest difference was between the irreligious (42–56) and the devout (29–68), but both of them come down on the same side. The highly educated (41–58) were not much less opposed than those who never went to college (35–63).

On **sex selection**, the 2004 GPPC poll indicated that 57 percent opposed using PGD for that purpose. Their 2002 survey registered 68–28 opposition, fairly consistent among social groups, slightly higher among women and college graduates.

On **GE food**, opinions even in the US, where the issue is much less prominent than in Europe, are split evenly—46 percent think it's safe, 46 percent unsafe, and 9 percent are not sure.[27] However, the same 2003 poll registered a whopping 92–6 majority in favor of labeling GE food. Would seeing such a label make you more likely to buy GE food? No, 6–55, with 37 percent saying it would make no difference.

A Pew survey taken at about the same time generally confirms these findings.[28] It also reported a clear ranking of people's discomfort with genetic modification, which was consistent in *order* (though not necessarily in absolute amount) across all subgroups—age, gender, education, and so on. On a scale of 1 to 10, people (known to this survey as "consumers") are "most comfortable with modifications of plants (6.1) . . . [then] genetic modifications of microbes (4.2), animals used for food (3.8), insects (3.6), followed by animals used for other purposes, including horses, cats, and dogs (2.3). Consumers are least comfortable with genetic modifications of humans (1.3)."[29]

One subset of the animal modification issues, considered next in some detail, leads to some intriguing data about why there is so little mainstream discussion about making this unpopular activity illegal—as most people would prefer.

■ ANIMAL CLONING ■

Animal cloning itself is not part of the subject matter of this book, but the polls about it are both interesting and relevant. First, here is the public's impressively consistent view of animal cloning, in summary form:[30]

POLL	DATE	APPROVE (%)	DISAPPROVE (%)	OTHER (%)
Gallup	05/04	32	64	4
Gallup	05/03	29	68	3
Gallup	05/02	29	66	5
ABC	08/01	37	59	4
Gallup	05/01	32	64	4
Time/CNN	02/01	29	67	4
Time/CNN	02/97	28	66	6

The 2002–4 Gallup polls asked whether animal cloning was "morally acceptable" or "morally wrong"; the 2001 version whether it "should or should not be allowed." ABC flat-out asked if it should be "legal" or "illegal"—and *illegal won* by what we'd call a landslide in electoral politics. The 1997 poll from Time/CNN also revealed that 56 percent would not eat cloned meat.

Fox News asked about some specific cases of cloning animals:

	DATE	APPROVE (%)	DISAPPROVE (%)	OTHER (%)
Endangered species	02/02	29	64	7
Endangered species	04/01	32	61	7
Livestock	02/02	23	71	6
Livestock	04/01	27	66	7
Pet dog or cat	02/02	12	84	4
Pet dog or cat	04/01	16	79	5

Other polls found similar numbers, which seem to show that situations which might be thought to be more acceptable are in fact less so—people definitely do not want pets to be cloned, for example, or even endangered species.

So why is animal cloning allowed?

Part of the reason is certainly that cloning laboratory animals—mice, especially—for research purposes can be very useful; it enables scientists to hold a number of variables essentially constant. Another is sheer scientific curiosity, which may eventually lead somewhere. But that only explains why some people *want* to do it, and begs the question of why they are *allowed* to.

The 2001 Gallup and ABC polls included some breakdowns of opinion by various social categories. Assuming that little has changed since then—and the overall numbers have not—then, despite the large overall majority against, there are actually significant groups that tend to *support* animal cloning. As the Gallup News Service summary noted in a subheadline,[31] "Majority of Higher-Educated and Higher-Income Americans Support Animal Cloning." Specifically:[32]

- 56 percent of those with postgraduate education
- 52 percent of those earning over $75,000
- 64 percent of those earning over $100,000

According to Gallup, a majority of those who say that religion is not very important in their lives are also in favor of animal cloning; this drops to 40 percent among those who say religion is "fairly" important, and to 22 percent for those to whom religion is "very" important. There is also a gender gap of substantial proportions, for women are strongly against animal cloning (71–25 in the ABC poll, 74 percent opposed in the Gallup) while men are evenly split, slightly in favor of it in the ABC and somewhat against it in the Gallup, both within the margin of error.

■ POLITICAL IMPLICATIONS ■

The limited data that is available, for example from the VCU, GPPC, and Pew surveys cited above, strongly suggests that the *trend* shown by the animal cloning data (but not necessarily the actual numbers) is true of other, related issues.

The 2001 ABC poll had some figures on human cloning that support this: Men were less opposed than women (82–16 against, versus 93–6); those earning over $100,000 were less opposed than those earning less than $25,000 (80–20, versus 92–8); and the irreligious were much less opposed than the evangelicals (77–22, compared with 95–3).[33]

The figures on "therapeutic" (that is, research) cloning were in fact remarkably similar to those on animal cloning, in the same ABC survey: Overall, people thought it should be *illegal* (63–33), but those making more than $100,000 a year thought it should be *legal* (55–44) as did those with "no religion" (53–46). A separate poll, two weeks earlier, on stem cell research demonstrated a similar trend, though with different figures, and added the datum that blacks were much less supportive than whites—which might also have something to do with income, since blacks on average earn less, and possibly with religious beliefs, too (these categories are never simple).[34]

So: Affluent, educated, white men tend to like animal cloning, especially if religion is not important to their lives. In other words, technocrats like animal cloning. They are also much less opposed to other kinds of biotech than the rest of us. Meaning no disrespect to any individual in Congress, they all earn six-figure salaries, they are overwhelmingly white and male and they are usually well-educated . . .

Of course, the politics is much more complicated than that. But Democrats in Congress, in particular, are mostly in favor of biotech, not for crass reasons of campaign finance and revolving-door job opportunities, but because they believe in it. It would take considerable pressure to make them vote against their instincts.

But most feminists, environmentalists, people of color, and progressives in general, even if (as seems likely) they oppose animal cloning, are not going to consider it a deal-breaker that will send them into the arms of the Republicans; it's just not that important to them. Meanwhile, the religious right has other priorities—the anti-abortion movement is much more important to *them* than animals are. Is it any surprise that there have been so few moves to make animal cloning illegal? Even though most people want a ban, it's not on the national agenda. It's no one's priority and most legislators don't support it.

California, however, may be leading the way on animal cloning. A bill to ban the sale within the state of cloned or genetically modified pets was introduced in February 2005 by a young, liberal environmentalist, prompted by a coalition of animal-rights and other activists. They are opposed by a company backed by the billionaire John Sperling (see **Chapter 10**), and supported by a large majority of the population.[35] Pet cloning may seem to be a trivial issue at first glance, but it reaches to important questions of principle, and it is not clear whether money or public opinion will prevail.

When it comes to Human GE, there has essentially been a stalemate for years (see **Chapter 12**). There is unquestionably a progressive opposition, as well as a religious one (see **Chapter 11**), but the conventional left-right single line continuum does not describe this very well or usefully. The opinion of the American public on genetic engineering, even when evidently certain, has not yet been reflected by the conventional political structure.

■ FURTHER READING ■

Free Documents from the Web

Matthew C. Nisbet, "Public Opinion About Stem Cell Research and Human Cloning," *Public Opinion Quarterly*, Spring 04, http://poq.oupjournals.org/cgi/content/full/68/1/131

Health Poll Search is an archive of public opinion questions on health issues, the result of a partnership between the Kaiser Family Foundation and the Roper Center for Public Opinion Research at the University of

Connecticut. You can search by topic or by key words, and also by polling organization, at www.kaisernetwork.org/health_poll/hpoll_index.cfm

Polling Report summarizes polls, and groups them into categories, including www.pollingreport.com/science.htm, www.pollingreport.com/health.htm and www.pollingreport.com/values.htm, which are all cross-linked. There is a subscription service ($95/year), which provides extras, including a newsletter, but plenty of data is available free, covering everything from political campaigns to the likelihood of Elvis being alive.

The Pew Initiative on Food and Biotechnology, pewagbiotech.org, claims to be impartial but is tightly connected with the establishment, with former Secretary of Agriculture Dan Glickman at the helm. It conducts annual surveys of "Public Sentiment About Genetically Modified Food," available from www.pewagbiotech.org/polls.

The Center for Genetics and Society has an excellent compilation of relevant polls from all round the world, at least up to 2003, at www.genetics-and-society.org/analysis/opinion/index.html.

The True Food Network maintains a list of polls about GE food at www.true foodnow.org/home_polls.html.

The Virginia Commonwealth University (VCU) Life Sciences surveys are available at www.vcu.edu/lifesci/centers/cen_lse_surveys.html.

■ ENDNOTES ■

1 National Science Board, *Science and Engineering Indicators 2004*

2 *2003 VCU Life Sciences Survey: Public Values Science But Concerned About Biotechnology*

3 "New Poll: Americans Oppose Destructive Embryo Research, Support Alternatives," Press Release, National Conference of Catholic Bishops, 06/08/01

4 Poll conducted by Caravan OCR International, 05/10–13/01

5 *The Wire*, Newsletter for the Eastern Region Biotechnology Initiative (UK), Nov. 1999

6 "Public Support For Controversial Technologies Could Increase If Applications Are Explained," MORI Summary, 09/08/99

7 www.truefoodnow.org/home_polls.html

8 See, for example, "Ten Reasons Why GE Foods Will Not Feed the World," at www.organicconsumers.org/ge/tenreasons.cfm.

9 CNN/*USA Today*/Gallup Poll, 11/14–16/03

10 Neil Munro, "Doctor Who?" *Washington Monthly* 11/02

11 "Public Awareness and Attitudes about Reproductive Genetic Technology," the Genetics and Public Policy Center (GPPC) with Princeton Survey Research Associates, 12/09/02

12 ABCNews.com, 03/29/01

13 Robin Marantz Henig, "Adapting to Our Own Engineering," *New York Times*, 12/17/02

14 Harris Poll conducted for *Parents* magazine, reported in the *San Francisco Chronicle*, 09/14/78

15 Gallup Poll, reported in the *San Francisco Chronicle*, 12/14/78

16 *Science*, 06/18/04

17 KRC poll, 06/04

18 GPPC, op. cit.

19 From Matthew C. Nisbet, "Public Opinion About Stem Cell Research and Human Cloning," *Public Opinion Quarterly*, Spring 04

20 International Communications Research poll, reported 08/26/04

21 University of Pennsylvania National Annenberg Election Survey, 07/30–08/05/04

22 Nisbet, op. cit.

23 The Genetics and Public Policy Center (GPPC, www.dnapolicy.org) issued a press release with these numbers on 05/03/04.

24 CNN/*USA Today* survey, conducted by Gallup on 01/23–25/03

25 Gregory Stock, *Redesigning Humans: Our Inevitable Genetic Future*, New York: Houghton Mifflin, 2002, p. 58

26 Darryl R. J. Macer, "Public Acceptance of Human Gene Therapy and Perceptions of Human Genetic Manipulation," *Human Gene Therapy*, 3 (1992), 511–8

27 ABC News poll, conducted 07/9–13/02

28 "Public Sentiment About Genetically Modified Food, September 2003 Update," Pew Initiative on Food and Biotechnology

29 ibid.

30 Kaiser/Roper, time-com and cnn.com

31 Joseph Carroll, the Gallup Organization, Poll Analysis, 06/07/01

32 ibid. and Dalia Sussman, "Majority Opposes Human Cloning," analysis at ABC News, 08/16/01

33 ibid., Sussman

34 Jesse F. Derris, "Stem-Cell Backing Holds at Six in 10," ABC News, 08/03/01

35 Yet another opinion poll with the same numbers, this one commissioned by the American Anti-Vivisection Society (AAVS), is linked from the resources page of NoPetCloning.org, home of Californians Against Pet Cloning. The countersite is YesPetCloning.org, also known as DefendPetCloning.org.

10

ADVOCATES AND ENABLERS:
THE PEOPLE BEHIND HUMAN GE

■

■

■ INTRODUCTION ■

Most people don't *want* to change their children's genes (see **Chapter 9**). Some do, however, and they are an increasingly vocal minority, who have both the contacts and the expertise to make their views part of the public discourse.

The pro-GE pitches supporters peddle vary in their details but share one salient characteristic: they are filled with sunny optimism. They are facets of a vision of technological utopia, and as such superficially appealing. The choices before us are framed as intelligence over stupidity, health over illness, beauty over ugliness, life over death. (Opponents, it should be noted, argue each point, vigorously.)

Further, advocates frequently claim that their concepts of the future represent the culmination of the drives that have made our society what it is, and as such are inevitable. Failing that, they tend to assert the libertarian view that government has no right to restrict research; they often drift all too close to "I wanna and you can't stop me." And the fallback position is, of course, medical: Developing Human GE will cure (or eliminate) disease.

All of these assertions are questionable at best, and that's being excessively polite. Some of the arguments and counterarguments have been made in **Chapter 3**, on **Cloning**, **Chapter 8**, on **Eugenics**, and indeed throughout this book; more are addressed specifically in **Chapter 11**, which focuses on reasons *not* to attempt Human GE.

The advocates, supporters, and enablers of Human GE are, however, more than able to speak for themselves. They often do: Some of the supporters are well-connected in the mainstream media and readily available for TV, while many more have extensive websites. Rather than risk mischaracterizing their opinions any further, this chapter serves as a brief survey of the intellectual landscape and a resource guide, spiced with some salutary examples of individuals to watch out for. They're not the worst, necessarily; none of them are stupid, most are pleasant, some very engaging, and all on at least some level sincere. They are here to give a flavor of the advocates' idiosyncratic and privileged universe.

■ SCIENTIFIC ADVOCATES ■

MOST MICROBIOLOGISTS, like most scientists in general, are *not* publicly active in politics; they stick to their lab benches. The little polling data that exists (see **Table 3.1** and **Box 4.3**) suggests that most of them *do* share at least some of the general public's ethical concerns: A large majority oppose creating human embryos just for research, and an overwhelming majority are against reproductive cloning. Very few indeed actively campaign either for or against genetic engineering of any kind, and of them only a minority concern themselves publicly with the Human GE debates.

That said, there certainly are scientists who are actively engaged in promoting Human GE, who work in reputable institutions and who have considerable standing in society. They include Nobel laureates, of whom the most prominent is the Grand Panjandrum of Molecular Biology, James Watson (see **Box 10.1**). Watson is of retirement age—he was born in 1928—but still very much available for public appearances.

10.1 THE THOUGHTS OF CHAIRMAN JIM

JAMES WATSON was one of the discoverers of the structure of DNA (see **Box 2.8**). Later, he ran the Cold Spring Harbor laboratory, was the founding head of the Human Genome Project, and has become one of the most outspoken promoters of germline human genetic engineering.

Watson won the Nobel Prize but has admitted that "the double helix was going to be found in the next year or two" and put his success down to "ambition."[1] As a young academic, he was "the most unpleasant human being I had ever met," according to E. O. Wilson.[2] He liked to see himself as "Honest Jim" and became notorious for statements such as:

- "People say it would be terrible if we made all girls pretty. I think it would be great."[3]
- "Whenever you interview fat people, you feel bad, because you know you're not going to hire them."[4]
- "Anyone who would hire an ecologist is out of his mind."[5]
- "It's not much fun being around dumb people."[6]
- "Take love . . . love doesn't come from God, so it's not the greatest gift of God but the greatest gift of our genes. . . . [If you lack a "love gene" then] as long as you've got a good brain, you can marry for money."[7]
- "Variations in DNA are responsible for why we are all different, our predisposition to disease, violence, sense of humor—even Francis Crick's laugh."[8]
- "Some people are going to have to have some guts and try germline therapy without completely knowing that it's going to work."[9]
- "I think our hope is to stay away from regulations and laws whenever possible."[10]
- "Scientists should proceed unhindered towards germline engineering."[11]
- "I think that the acceptance of genetic enhancement will most likely come through efforts to prevent disease."[12]

He also extolls "the freedom to have your brain work right. As distinct from most people, whose brains don't work right."[13] And we are left with the distinct impression that Watson does know intimately at least one exception to that generalization.

Watson always did cultivate a reputation for outspoken eccentricity. He had a habit of ruffling his hair and untying his shoelaces before going into meetings with funders.[14] He himself has speculated that he had the "root form" of "a gene associated with violence" but kept it under control because he came from a "good family."[15] Nevertheless,[16] "You know, when I'm in a room, and I hear shit, after a while the word 'shit' is going to come out. You just can't take it anymore. Now that's, hmmm, a predictable response. It's bound to come out. I think to myself, maybe I'll sit through nonsense and not say it. But . . ."

This appears to be a mild form of coprolalia, "the uncontrolled, often excessive use of obscene or scatological language" (commonly referred to as being obnoxious).[17] According to his former partner (see **Box 2.8**), he also has some minor problems of physical coordination. One might even say that his brain doesn't work right. That's no more unfair than Watson routinely is about those who disagree with him. What he does is say the unsayable. He reveals the prejudices behind much of the gene project he did so much to advance.

Watson has been berated for his flaws many times, most devastatingly perhaps in a 2003 review by Susan Lindee, published in *Science*, of what she called his "public promotion of genomics, DNA: The Secret of Life."[18] Noting that "throughout his account, Watson is unconstrained by either evidence or logic," she nails him for "vintage eugenics" (which he denies; but see **Chapter 8**) and stresses that "health care expenditures should reflect human needs, rather than potential corporate profits." In conclusion, she agrees sarcastically that the Human Genome project revealed "what makes us human"—only in her view, it reveals "our tendency to elevate what we craft into the realm of neutral, absolute truth, and make manifest our vulnerability to propaganda. Watson has been the genome project's marketing director and prime salesman. His latest promotional brochure is not worth anyone's time."

The key gathering of academic Human GE proponents was in 1998 at the University of California, Los Angeles (UCLA), where Gregory Stock (see **Box 10.2**) and others organized a Symposium on Engineering the Human Germline. Watson described this as "the first gathering where people have talked openly about germline engineering," which he regarded as a very good thing. Certainly, it helped to legitimize the discussion, and encouraged networking among the advocates.

10.2

GREGORY STOCK: HAVE SPEECH, WILL TRAVEL

GREGORY STOCK was Director of the Program on Medicine, Technology and Society at the School of Public Health at the University of California, Los Angeles (UCLA) for half a dozen years, beginning in 1997. He's still a visiting professor at UCLA, and uses that as a base for a plethora of public appearances. His website lists four different "speaking agents" including one in Canada and one in the UK, as well as a "literary agent" and long lists of his publications and appearances.[19]

The Wall Street Journal called Stock "a rarity: a serious scientist who publicly supports cloning humans."[20] He has been described as "a bridge between fringe post-human activists and the mainstream scientific community."[21] In that capacity, he—and his UCLA credentials—have been welcomed at conferences organized by the Extropy Institute and the World Transhumanist Association (WTA; see below).

His website lists a 2003 appearance as having taken place at Yale (which is true) but not as a WTA event, which it was. This may indicate a certain ambivalence about the identification, but there is no doubt where his sympathies lie. He debated George Annas of the Boston University Health Law program (who thoroughly disapproves of Human GE), on the topic, "Should Humans Welcome or Resist Becoming Posthuman?"

Stock, naturally, took the "Welcome" side: "I don't care about the species, I care about individual people. . . . [We should] not just accept, but embrace [the new technologies] because they're filled with promise and because we can."[22]

10.2 In 1998, Stock was one of the principal organizers of a landmark event: a Symposium at UCLA on Engineering the Human Germline, which gathered together many of the most significant players in the science, including the ubiquitous Watson. The proceedings were developed into a book of the same name, edited by Stock and his UCLA colleague, John Campbell and published in 2000.

Engineering the Human Germline stands as a foundational text of the serious, semi-academic pro-GE approach. Two years later, Stock published a popular book along the same lines, Redesigning Humans. It's a readable and superficially convincing account, and all the more dangerous for that.

Stock has a doctorate in biophysics from Johns Hopkins, an MBA from the Harvard Business School, a million dollars from his earlier Book of Questions and its sequels, and a real talent for publicity; he's said to have appeared on over 1,000 radio shows. He is bright, glib, funny, and a tireless proselytizer for his point of view. If he ever gets round to debating Lori Andrews (see **Chapter 5** for quotes from both of them), the proceedings should be simulcast on C-SPAN and the Comedy Channel.

The germline debate has largely been overshadowed by the media focus on human cloning and the vexed question of embryonic stem cell research. To some extent this is happenstance; the eccentrics who want to commit cloning now (discussed in **Chapter 3**) are *not* in the mainstream. The first reaction to them from these advocates of human GE seems to have been concern that the "cowboys" would derail the enterprise by provoking restrictive legislation.

On second thoughts, however, they realized that these efforts, handled tactfully, could provide convenient cover for the longer-term agenda: By agreeing with the majority that reproductive cloning should *not* be attempted (even if only for the moment, and only on safety grounds), they could begin to build alliances and lay the groundwork for the acceptance of technologies that they accept are not yet ready for implementation. Especially if they could find a medical excuse. And that is exactly what happened.

■ THE EMBRYONIC STEM CELL LOBBY ■

NEITHER EMBRYONIC STEM CELLS nor even cloning tech-nology need necessarily be connected with inheritable—germline—Human GE. Human GE is, however, almost inconceivable without them (see **Chapter 2**). It's therefore difficult to separate the medical from the enhancement-related motivations of people advocating, and performing, embryonic stem cell research. They all *claim* medical justification, and few of them are willing to go on record as supporting either germline intervention or cloning.

A key exception is another Nobel laureate, of the same gen-eration as Watson: **Paul Berg**, who won his for work with recombinant DNA. In the 1970s, Berg took a lead in setting up a very temporary—and entirely *voluntary*—moratorium on work in his specialty (see **Chapter 12**), for which he has earned an otherwise unwarranted reputation as a moderate. In fact, that is also as far as he would like to go in dealing with human repro-ductive cloning—a voluntary agreement among scientists. He has consistently considered legislation to be,[23] "the wrong way to deal with advances in science. We need a mechanism that is more flexible. . . . Legislation closes doors that you may want to open six months from now."

When California was considering its ban on reproductive cloning, which was coupled with an endorsement of research cloning, Berg acknowledged in his testimony that he opposed the ban and accepted it only as a compromise position.[24] His view, which he also expressed to a US Senate committee, was that reproductive cloning should be legal when (if) animal experiments provided more reliable results. He advocated "reviewing the statute periodically, perhaps every ten years, to determine if the judgments made today remain valid in the light of new scientific information."[25] Note: By strong implica-tion, Berg considers ethical considerations irrelevant.

Berg's prime expressed goal has been to avoid restrictions on research cloning. In this, he has been joined most prominently by **Irving Weissman**, Stanford professor and multimillionaire entrepreneur.[26] Weissman chaired the 2002 National Academy of Sciences committee that recommended banning reproductive cloning, on safety grounds of course. He made a substantial fortune from cofounding several companies, including SyStemix, netting a reported $20 million when the latter was sold to Sandoz, which became Novartis; SyStemix was set up to exploit adult stem cells for gene therapy, and Novartis took a major hit when that didn't work.[27] Adult stem cells have been Weissman's entrepreneurial speciality—another company he cofounded is StemCells Inc.—but he has always advocated cloning research and was a prime spokesman for the 2004 California initiative that funded it (see **Chapter 12**). In his TV commercials for that, he carefully avoiding "promising" cures; that, he insisted to a reporter from *Science*, would be "just not right."[28] Instead, he said,[29] "The chances for diseases to be cured by stem cell research are high, but only if we start. If the promise of stem cell research comes true, we can hope for a single treatment with the right stem cells to cure diseases every family has." Shares in StemCells Inc. rose from a 12-month low of $1.24 to a 12-month high of $4.87 just before that election, while those commercials were running; Weissman held options that became valuable at $5.25.[30] The dictionary definition of "disingenuous" is "not straightforward or candid; crafty."[31]

Other key practitioners of embryonic stem cell research include Douglas Melton of Harvard, Lawrence Goldstein of the University of California San Diego School of Medicine, John Gearhart of Johns Hopkins, and James Thompson of the University of Wisconsin, who was the first to isolate human embryonic stem cells. All of them are regularly quoted in the popular press; most of them generally avoid addressing the broader questions of germline intervention. That's normal.

■ PASSIVE ENABLERS ■

MOST WORKING SCIENTISTS presumably sympathize with Tim Tully, of the Cold Spring Harbor Laboratory, who works on the genetic basis for memory. He has admitted that there is "a potential here for serious abuse" and that there are important questions about, for example, the consequences of a possible drug to block memory of trauma. His response,[32] "That's a tough one. Thank God I don't have to answer it. I just play with flies."

There is of course a longstanding argument about the responsibility of scientists for the application of their discoveries. It could certainly be said that all those working on genetic technologies are in some sense *enabling* their possible misuse; but given the possibility of discoveries that may lead to medical benefits, that's unfair. Basic research is certainly important.

It does, however, appear to be true that some scientists working on uncontroversial areas of research have been pressured to support those working on areas that are more likely to raise ethical concerns. Dr. Helen Blau of Stanford, a distinguished researcher, has been quoted as saying that by speaking out, "I could get in a very big mess [with] everybody."[33]

It's safer, then, to take a position of live-and-let-live. Which fits very well with the strongly libertarian bent of much of the Human GE advocacy.

■ LIBERTARIANS ■

THE FORMAL US Libertarian Party came out in support of human reproductive cloning as early as February 1997, immediately after the announcement of the first cloned mammal, calling cloning (whether of animals or humans) one of the "most exciting and important scientific breakthroughs of the 20th Century." They explicitly linked it with reproductive rights,[34] "Politicians should not have veto power over the creation of new life—especially human life. That's why the Libertarian Party supports reproductive freedom of choice for Americans—whether they choose to reproduce using the traditional method, or artificial insemination, or in-vitro fertilization, or cloning."

The political party itself is small—their Presidential candidate received around 400,000 votes in 2004, somewhat behind Ralph Nader—but their sympathizers and fellow travelers are quite influential. They include:

- ▶ **Reason** magazine, whose science editor, **Ronald Bailey**, writes and speaks extensively against what he calls the "reckless conservatism" of those who would place any limits on research[35]
- ▶ **Virginia Postrel**, former editor of *Reason*, author of *The Future and Its Enemies* and *The Substance of Style*
- ▶ **The Ayn Rand Institute**, whose spokesperson Alex Epstein says that parents "should have the right to play God with the genetic makeup of their children," and that cloning researchers should not be "shackled by their government" but "celebrated as the heroes they are."[36]

A rich vein of entitlement runs through much libertarian thinking, and it is echoed by many Human GE advocates who may not consider themselves to be libertarians. Few ever mention the social justice implications of their positions, especially if they are considered in class terms: the idea that GE will help "us" to maintain our superiority over "them." Those that do, however, seem untroubled by the prospect, or at best regard it as inevitable and therefore not a concern; see **Box 10.3** for the most egregious example.

10.3 LEE SILVER, REMAKING SOCIETY

PROFESSOR LEE SILVER, of Princeton, is a mouse geneticist with a talent for entertainingly glib writing. The most notorious section of his book *Remaking Eden* discusses the notion that Human GE will lead eventually to the separation of people into two distinct species. This book has had an impact—in fact, it has single-handedly converted many people to a point of view quite the opposite of his own.

Silver's extreme scenario is that those who can afford genetic "improvements" will have them, and become the ruling class, whom he calls the "GenRich." All menial work will be done by the "Naturals." Eventually,[37] "the GenRich class and the Natural class will become the GenRich humans and the Natural humans—entirely separate species with no ability to crossbreed, and with as much romantic interest in each other as a current human would have for a chimpanzee." In the bad old days, owners would sexually exploit their slaves. In this imagined future, they won't even want to. This is progress?

Silver presents his horrifying predictions more in sorrow than in anger, claiming they are inevitable. But he doesn't seem to have learned a thing since the mid-1990s. That's when *Remaking Eden* was written (it was published in 1997), and it reflects both the "geno-centrism" and the libertarian boom-time approach of the era, with its impossible dreams of stock market wealth for all. He's sticking to it, insisting in October 2004 that "biotechnology is using living things in design, to design things that people can use" and, over and over again, that cloning is inevitable.[38]

It's a strange faith, this doctrinal belief in the inevitability of degradation through science. Strange, and quite irrational.

■ THE EXTROPIANS ■

OUT OF A small but complicated libertarian/futurist/techno-freak culture of the 1980s and 1990s developed a number of intense and, to most people, intensely weird groups devoted to remaking, reenvisaging, and generally redesigning people.

Perhaps the most libertarian of these is the **Extropy Institute**, although even they show signs of backing off in the direction of democratic society. Their FAQ used to include a reference to opposing "the redistribution of wealth through forcible taxation and regulation of commerce" but that phrase has been removed.[39] Nor does the document any longer call for the abolition of the Food and Drug Administration (FDA) on the grounds that "any government monitoring of drugs at all is unnecessary." Still, that's still definitely where they're coming from.

Max More, the former Max O'Connor (self-reinvention is, after all, the point here; he's an emigrant Brit with an Oxford degree), and the even more delightfully renamed T. O. Morrow (formerly Tom Bell) cofounded *Extropy* magazine in 1988. In his keynote address at the 1999 Annual Convention, More pronounced:[40]

> We have decided that it is time to amend the human constitution. . . . Through genetic alterations, cellular manipulations, synthetic organs, and any necessary means, we will endow ourselves with enduring vitality and remove our expiration date. . . . [W]e will seek complete choice of our bodily form and function, refining and augmenting our physical and intellectual abilities beyond those of any human in history.

Sometimes this stuff reads like a boys' club satire. It *is* basically a boys' club: According to a 2002 survey, 80 percent of Extropians are male and more than half are under thirty.[41] The community evolved out of what a close observer has characterized as "white, male, affluent, American Internet culture." There are, however, women involved, notably More's wife, Natasha Vita-More (the former Nancie Clark), bodybuilder, artist, and now president. She was once an elected Green official, but left the party because they were, of all things, "too far left and too neurotically geared toward environmentalism."[42] Who'd have thought?

Despite all this silliness, the Extropians deserve mention because their elaborate and apparently well-funded conferences have attracted as speakers some genuine players in the human genetics research community, including:

- Gregory Stock (twice)—he's on the Council of Advisors
- John Campbell, the coeditor of *Engineering the Human Germline*
- Calvin Harley, chief scientific officer at Geron
- Professor Cynthia Kenyon of the University of California, San Francisco, a prominent researcher into aging

Other members of the Council of Advisors ("Experts of Knowledge") include Marvin Minsky of MIT, Ray Kurzweil (inventor and futurist), and Bart Kosko, the fuzzy-logic maven. These are very bright people. Don't be fooled by their veneer of idiocy. Beneath the surface craziness is a deep and potentially dangerous foolishness. And below that? Who wants to know?

▪ THE WORLD TRANSHUMANIST ASSOCIATION ▪

THERE ARE A plethora of other groups in the fractured circle of trans- and post-humanism. **The World Transhumanist Association (WTA)** is probably the largest and best organized. They see themselves as the moderate, "liberal democratic" wing of the post-human movement. The group developed at least partly in reaction to the Extropians, because that brand of "anarcho-capitalist orthodoxy" did not sit well with many Europeans. That's according to the most authoritative historian of the movement, WTA secretary James Hughes, a relative beacon of reason in this eccentric universe.[43]

Certainly the WTA is more sympathetic to the possibility of taking public policy measures to avert catastrophe, where the

Extropians tend to rely on pure market forces. The WTA leans to utilitarianism, the Extropians to pure individualism. The groups are, however, formally affiliated.

The WTA has rapidly become a significant body, with about three thousand members in ninety countries. Their *Transvision 2003* and *2004 Conferences* featured 53 and 46 speakers, respectively (many of them were at both).[44] The WTA is still, however, at a stage where it boasts of an attendance of 150; the report in *Reason* estimated 125.[45] It seems that at least a third of the attendees were on the platform at one time or another.

■ OTHER FRINGE GROUPS ■

THE LOOSE WEB-RING CULTURE in which much of the communication about post-humans takes place rather invites invasion by neo-Nazis and other racists and open eugenicists. Inevitably, this has happened. The WTA is firmly anti-racist and anti-fascist, but free speech concerns have caused some internal arguments; they seem to have been resolved by excluding racists from a newer web-ring.

Confusingly, there are two groups with very similar names—**Prometheism** and the **Prometheans**. The latter was founded by Phoenix, aka Colin Patrick Barth, who has "surpassed and synthesized . . . environmentalism, technologism, republicanism, conservatism, libertarianism, altruism, Objectivism, egoism, moralism, Platonism, statism," and just about everything else.[46] These Prometheans are stongly anti-eugenic, though they do "support preserving the option of voluntary genetic modification on an individual basis."[47]

Prometheism, however, is emphatically in favor of eugenics, specifically neo-eugenics. You can join their "religion" by swearing to "purposefully direct the creation of a new post-human species. . . ."[48] The website says it was founded by Dr.

Matt Nuenke in 2000. He may or may not be the "Marcus Eugenicus" who was outed as a Nazi by James Hughes; it's hard to care. Their proclamation also includes vague threats of inter-neohuman-species warfare, which may be meant to be purely intellectual, and thus metaphorical, or perhaps not. Frankly, the main justification for including them (him?) is comic relief. But that's important.

■ A LOOSE AFFILIATION OF MILLIONAIRES AND BILLIONAIRES ■

SERIOUS MONEY, however, is no laughing matter. There are several individuals with both the cash and the inclination to make a significant difference in the development of GE. John Sperling (*next*) already has: He paid for the cloning of cats. In significant if sometimes more indirect ways, John Templeton (*below*) has, too, and this may be just the start: Private financing could lead to really terrible outcomes, in the absence of serious national regulation.

At first sight, this may seem unlikely. Overall, the National Institutes of Health (NIH) is by far the biggest source of funding for research in the field; its 2005 budget is somewhere just north of $28,000,000,000 ($28 billion; it's worth looking at the zeros occasionally).[49] That's not all spent on Human GE, but still it's a big stick with which to enforce guidelines (see also **Chapter 12**), and absolutely dwarfs charitable donations.

All healthcare-related foundation grants in 2002 put together amounted to a little under $3 billion.[50] Of that, about one-sixth came from the William and Melinda Gates Foundation and a little less from the Robert Wood Johnson Foundation (no other single entity supplied more than about 5 percent). Obviously, those large foundations have the potential to affect research priorities quite directly, but in practice neither of them shows any particular interest in Human GE. Smaller foundations can, however, have a quite disproportionate effect, by targeting underfunded niches. Like pet cloning, for example.

■ JOHN SPERLING: CATS AND PEOPLE ■

JOHN SPERLING made his fortune, generally estimated at around $3 billion, from the University of Phoenix, a for-profit adult education institution, which he founded in 1976. It has been accused of being a "diploma mill" and "McUniversity" but it certainly found a niche.[51] The holding company went public in 1994, when he was seventy-three, and his wealth exploded.

He's used the windfall in part idiosyncratically, giving $3.4 million to sponsor medical marijuana initiatives and spending a couple of million on political ads with a leftist, populist bent in 2004.[52] He's a self-made man—literally born in a log cabin in the Ozarks—and tweaking the authorities' noses seems to come naturally to him.[53]

Another project was to clone his dog, Missy, which Texas A&M failed to do. But the team did succeed in cloning a cat. Ever alert to commercial possibilities, Sperling backed his PR person, Lou Hawthorne, in setting up Genetic Savings and Clone, which sells pet cloning and "gene banking" services.[54] Hawthorne claims that the cat division will be profitable in 2005, dogs in 2006 (if they succeed in cloning a dog), and the company as a whole in 2007.[55] He envisages a multi-billion-dollar industry.

Sperling's real interest as he ages, however, has been in longevity research; immortality, if you will. He owns several biotech companies through Exeter Life Sciences, and funds the Kronos Longevity Research Institute, as well as the associated Kronos clinics. His attitude to Human GE is simple,[56] "I am 100 percent for human enhancement!" He is planning on leaving the full $3 billion to a foundation dedicated to trying "to develop cell therapies for regenerative medicine, possibly pursue therapeutic cloning technology on human cells, and maybe, someday, experiment with genetic engineering." Such an endowment may be able to do a lot of good, but the potential for abuse is horrifying. With that kind of money, NIH guidelines become irrelevant; what is needed is legislation.

But you don't need to spend anywhere near a billion dollars to make a big difference.

■ SIR JOHN TEMPLETON: SCIENCE AND RELIGION ■

THE TEMPLETON FOUNDATION is the instrument of Sir John Templeton, who made his money as an investor, to promote his idiosyncratic theological and social ideas. Born in 1912, he was still vigorous in late 2004, and especially interested in what he sees as the interface between science and religion, particularly in subjects such as the efficacy of prayer in self-healing and the potential of science to "perfect" the human body, including the possibility of corporeal immortality.[57] His relatively modest foundation provides a case study in how to change the discourse in a chosen field.

The Foundation controls close to $300 million (one-tenth of Sperling's, one hundredth of Gates's), and disburses something over $20 million a year.[58] It conducts some activities with great publicity, notably for the annual million-dollar-plus Templeton Prize for Progress in Religion. It also raises its own profile and that of its subjects of interest by making several hundred mostly modest donations every year to major universities, including Stanford, Duke, Emory, Georgetown, Harvard, Princeton, UC Berkeley, USC, Vanderbilt, Yale, Oxford (where Sir John, who now holds British citizenship though he was born American and lives in the Bahamas, endowed Templeton College), Cambridge, and London.

Most of its funds, however, are channeled through a small

number of organizations that derive the vast majority of their income from this source. In effect, it funds a few theologians and bioethicists who agree with his views, provides them with the legitimacy of financially viable institutions that can, for example, hold prestigious conferences, and encourages them to promote their opinions as if they were widely held.

Many scientists and academics are very hostile to the Foundation (one has called it "a hideous, evil organization") because of its deliberate attempt to influence the teaching of both science and philosophy.[59] The Prizes it sponsors are thought to have an undue influence on course selection, and more broadly there is concern that overemphasizing the commonality of science and religion "is dangerous—especially when the driving force behind the effort is not the strength of ideas, but one man's money," in the words of Lawrence M. Krauss, chairman of the physics department at Case Western Reserve University.[60]

■ THE BUSINESS OF LONGEVITY ■

LARRY ELLISON is another billionaire with interests in immortality, though he is a generation younger (born in 1944, he may be hearing the Grim Reaper tiptoeing up). His fortune has dropped from the heady days of 2000 when he had $58 billion and rivaled Gates, but in 2004 it was still a healthy $18.7 billion, according to *Forbes*.[61] He has interests in several biotechnology companies, and he too has a foundation—the Ellison Medical Foundation—focused on research about aging and infectious diseases.

It's a notable player, giving away some $20 million in 2002, and entirely reputable. Executive Director Dr. Richard Sprott has a clear view of some of the potential pitfalls in the longevity field:[62] "It is probably the world's most fertile field for snake oil right now. I'm concerned that people are getting sold a bill of goods: not just damaging to their pocketbooks, but much of it is dangerous. . . . Almost all scientists (in the age research field) are involved in the private sector to make a profit." In part, Sprott implies, this is due to a kind of naïveté: "Biologists have never been terribly concerned about the consequences of what they do. Many of them are doing it because the problem and challenge interests them."

But there is a demand. Some people actually have ambitions of abolishing death, and effectively treat life as a terminal disease—a regime that includes 250 dietary supplements a day is less alluring in the specific than the concept might be in the abstract.[63] Another person who wants to live forever, a Doctor of Philosophy, was said to be "somewhat hurt" when a reporter pointed out that he did indeed look his late-forties age rather than the twenties he felt.[64] "It's the inside that matters," was his rather lame rejoinder.

Among the serious scientists to watch in this field is Cynthia Kenyon of the University of California at San Francisco (UCSF), who has manipulated the genes of tiny little roundworms, *C. elegans*, and increased their lifespan by up to a factor of six.[65] Her company—of *course* she has one—is Elixir Pharmaceuticals.[66]

■ CRYONICS AND OTHER BUSINESSES ■

THE SUPPORTERS OF Human GE often represent a kind of individualistic idealism. They may be selfish, in the sense of wanting to benefit directly themselves (especially those interested in immortality), and some may have an eye on stock options and the like, but they do not generally represent enormous commercial forces at present. The economic clout involved is not on the scale of the agribusiness interests driving Food GE. *Forbes* summed up the financial prospects at the end of 2002,[67] "Being struck by lightning is very rare. Cloning a human being is even rarer. But actually making money off this kind of *Star Trek* technology may be the rarest feat of all."

The broad biotech industry is hedging its bets, however.

Pharmaceutical companies have their eyes open for opportunities, and the Biotechnology Industry Organization (BIO) consistently lobbies against effective regulation, particularly of cloning (see **Chapter 12**).

Some biotech companies are involved, aside from the side projects of university scientists. Two of the most significant, Geron and Advanced Cell Technology (ACT) were founded or owned by Michael West (see **Box 10.4**). Another is the Canadian Chromos Corporation, which is developing artificial human chromosomes, a technology that could have profound implications; that's a way to make a new species.[68]

For those who fear that their own life may end before the dreams of immortality and enhancement come true, there is always cryonics. Many of the Extropians and other trans- or post-humanists are reported to have contracted with Alcor Life Extension Foundation or other cryonics firms to have their full bodies or their heads frozen upon their death, in hopes of eventual reanimation.

This little niche industry got a publicity boost when the son of baseball great Ted Williams had him frozen, and got sued by other family members, who finally ran out of money and dropped the case.[69] The brouhaha was well summarized in the *Boston Globe* headline: "Williams Dispute Piques Interest in Cryonics, Except Among Scientists."[70] On a purely anecdotal basis, however, it was gratifying to hear how many sports talk radio callers thought the whole mess absurd.

10.4 MICHAEL WEST, FOREVER DREAMING

MICHAEL WEST is a true believer. He is mild-mannered and soft-spoken, with an innocent-scientist shtick that makes him easy to underestimate, but he has already spent tens of millions of dollars of other people's money (and some of his own) pursuing the dream of immortality.

West's background is that of a used car salesman—the family business was leasing cars and trucks—and Christian fundamentalist. He attended

10.4

Creationist colleges, and even participated in anti-abortion demonstrations, before being convinced that the "evolutionary viewpoint" was correct.[71] He evolved ("a bit like the Apostle Paul in reverse") into a science entrepreneur with a mission, which he has described as a "religious quest," to solve the mysteries of aging and mortality.

Geron

West sold the family business in the early 1980s, but he retained (or set up; accounts vary) a shell company which he named Geron, from the Greek for "old man" (as in "gerontology"). Armed with a doctorate in cell biology, he raised $250,000 seed money from a few eccentric, elderly rich folks, and dropped out of medical school in Dallas.[72] He hit the big time with a showstopping 1992 presentation to venture capitalists in California that raised $8 million in an afternoon.

Geron soon moved to California, ending up in Menlo Park, a little south of San Francisco, and collected a star-studded Scientific Advisory Board, including most of the leading researchers in the field, not to mention—it's a small world—James Watson. The advisors got $35,000 a year each for up to thirty-five days consulting and, more important, lending their reputations to this start-up.[73]

The company eventually burned through at least $40 million, collected a bunch of patents, and went public in 1996, which netted West personally some $2 million.[74] Geron financed the research that led to the first isolation of human embryonic stem cells, at the University of Wisconsin; later, it bought the company that cloned the first sheep. Even critics admitted that Geron energized the telomere field (a specific focus of anti-aging work), but it was accused of "unseemly aggressiveness" and,[75] "hiring on all the important [telomere] research players in academia, securing their intellectual property, and then consuming their research for the sake of the corporate agenda."

The details of West's leaving Geron in 1998 are shrouded by nondisclosure agreements, but his departure was not amicable. It seems that he was forced out of the corporation he founded, possibly because his evolving ambitions and visionary style did not sit well with the more orderly and product-oriented processes of a public company.

10.4 ACT

Advanced Cell Technology (ACT) was a small Massachusetts outfit working on cloning, principally of chicken, later cattle. Since West was by now full-bore on the clone-yourself story (see **Chapter 4**), this suited him perfectly, and he bought a controlling interest for about $1 million.[76] ACT rapidly developed a terrible reputation for conducting press release science and pulling stunts, such as:[77]

- Announcing in October 2000 that they had contracted to clone a newly extinct Spanish mountain goat, or bucardo, whose tissue had been preserved (the last, a female, was hit by a falling tree in 1999). No success has since been reported, and with whom would she breed anyway?

- Announcing in January 2001 that they had successfully cloned an endangered species, a gaur (a type of Asian ox), using a cow egg and a cow surrogate. The specimen died within two days, which ACT conveniently claimed was unrelated to its cloning.

- Announcing in November 2001 that they had cloned a human embryo, although it failed to develop beyond the six-cell stage, if that. The news of what Rudolf Jaenisch, an MIT cloning expert, called "this ludicrous, outrageous, failed experiment" was published in a new and until-then obscure on-line journal, simultaneously with a feature West and his colleagues wrote in *Scientific American* and a cover story in *U.S. News & World Report*.[78] It led to the resignation of at least three advisory board members of the journal, all of whom had been blindsided.[79]

At the time of the human embryo claim, ACT was "trying to raise $15 million to $20 million" but its then Chief Operating Officer, Robert Saglio, denied any connection.[80] The company has frequently been reported to have trouble making payroll, and it was unable to retain its chief cloning expert, Dr. Jose Cibelli, when Michigan State made him a better offer in late 2002.[81] Its website in November 2004 listed no scientific publications less than two years old, and only one press release in

10.4 the previous eighteen months; that described a setback in a patent infringement lawsuit brought by . . . Geron.[82]

California, Again

Even before California passed its stem cell research proposition in November 2004 (see **Chapter 12**), West had relocated back to the San Francisco Bay Area. The company was expected to follow. After all, it promised to rain money, and they had to hold their bucket out.

▪ BIOETHICS ▪

IN THEIR EFFORTS to encourage public acceptance of Human GE, supporters are assisted by a number of people who do not necessarily actively or openly endorse them. These include some artists and writers, as well as editors of mainstream media, but the most important contingent is from the bioethics community.

The study of bioethics certainly sounds like a good idea, and its practitioners have carved out a niche, both intellectual and—to be blunt—economic within the landscape of Human GE as well as related fields. The most prominent practitioner to the general public is probably Art Caplan (see **Box 10.5**), but they turn up all over the place. There's hardly a biotech company that doesn't have an ethics committee, for a start.

10.5 ARTHUR CAPLAN: A VERY PROFESSIONAL BIOETHICIST

ARTHUR CAPLAN runs the Center for Bioethics at the University of Pennsylvania, the largest university-based program in the country. He's the Emmanuel and Robert Hart Professor of Bioethics, and also, among many other appointments and activities:[83]

- ▪ a board member of Celera Genomics
- ▪ a member of DuPont's biotechnology advisory panel

10.5

- the former Chair of the Advisory Committee to the United Nations on Human Cloning
- a fellow of the Hastings Center and the American Association for the Advancement of Science (AAAS)
- the author or editor of 25 books and literally hundreds of academic papers
- a regular columnist for MSNBC.com
- a regular contributor to the New York Times, Washington Post, Philadelphia Inquirer, NPR, CNN, and "many other media outlets"

The Washington Speakers Bureau describes him, accurately, as "the media's go-to-guy [on] issues related to science and ethics" and charges $15–25,000 for an appearance.[84] He is the most visible bioethicist around, always available for a quote on the issue of the day.

Caplan's genial facility and undoubted smarts have enabled him to survive some difficult situations. He has criticized some ethical advisory panels for providing rationalization rather than guidance, while still taking part in others, and admitting that he could be seen as implicated with policies to which he objects.[85] He makes $20,000 a year consulting for industry (which some bioethicists object to strongly), and turns it over to the Center.[86] Even more controversially, he was connected with the Jesse Gelsinger tragedy (see **Chapter 6**), though in the end he was dropped from the related lawsuit.

What Caplan does exemplify is the tendency for bioethicists to *enable* even when they do not actively *advocate* Human GE. For example, Caplan and two of his colleagues wrote an article in the British Medical Journal titled "What Is Immoral About Eugenics?" which concluded that,[87] "No moral principle seems to provide sufficient reason to condemn individual eugenic goals."

And yet he is aware of the issue of "fundamental social inequalities." He and his coauthors are willing to assert that "force and coercion, compulsion and threat have no place in procreative choice" but seemingly unwilling to consider the coercive effect of advertising and manipulated social expectations. Crucially, he and his colleagues seem to value the rights of the individual over the rights of the community so routinely

10.5 that, even if "individual decisions can have negative collective consequences," such consequences don't give the community any "moral principle" by which to regulate the actions of the individual. If this isn't
advocacy, it's certainly enabling.

Caplan also predicted in 2000 that,[88] "making babies sexually will be
rare. . . . In a competitive market society, people are going to want to give
their kids an edge. They'll slowly get used to the idea that a genetic
edge is not greatly different from an environmental edge." That is certainly an odd position for a bioethicist to take: It's inevitable, so it's OK?
Situational ethics are taken to a remarkably circular extreme when the situation itself is invented and then the ethics derived from that imagined
future are applied to the present, so as to create that particular future.

What bioethicists *do*, however, has become increasingly controversial. Jim Watson—yes, him again—gave the field an enormous boost by allocating 3 percent of the Human Genome
Project budget to the ethical, legal, and social issues (ELSI)
involved in the project. For this, he still attracts undeserved
praise. In fact, according to Lori Andrews, who was on the
original ELSI committee, Watson later said that,[89] "I wanted a
group that would talk and talk and never get anything done. And
if they did do something, I wanted them to get it wrong. I
wanted as its head Shirley Temple Black."

The slur on a former child star who became a professional
diplomat (Ambassador to Ghana and Czechoslovakia and Representative to the United Nations) is typical, but so is the inci-
sive honesty. Ethics committees all too often do "investigate and
pronounce on controversial technologies only after they have
been developed."[90] Which means they are engaged in locking
the barn door well after the horse has left, and frequently in
denying the existence of the horse in the first place, no matter
what smell lingers. Those who discuss these issues all the time
get all too used to thinking the unthinkable and show an alarm-
ing tendency to justify the unjustifiable. A good rule of thumb
on ethics is: Don't trust the experts.

■ VIRTUAL ORGANIZING ■

ONE FINAL NOTE concerns the Internet. The enthusiasts for human GE are overwhelmingly technophiles and were early adopters of the Web. Almost everyone mentioned in this chapter has a website, which is no longer unusual; that most of them did in the 1990s however, is worth mentioning.

Indeed, Randolfe Wicker established himself as an authority on human cloning with an almost entirely "virtual" campaign of self-promotion—he got in early with a couple of websites and now the networks call him for comment.[91] To be fair, Wicker's generally skeptical about the charlatans, but he remains an active proponent of cloning in principle.

He serves, however, as a useful reminder, if one is still needed, that a vibrant Web presence is no guarantee of real-world support. Which doesn't mean there isn't a lot of good information out there . . .

■ FURTHER READING ■

Free Documents from the Web

Carl Elliott, "Humanity 2.0," *Wilson Quarterly*, Autumn 2003, is an overview of Transhumanism by a skeptical philosopher, available at www.mindfully.org/Technology/2003/Transhumanist-Humanity1sep03.htm.

Ralph Brave, "James Watson Wants to Build a Better Human," *Alternet*, 05/29/03, surveys the academic advocates of Human GE through the prism of Watson's eccentricities, at, www.alternet.org/story/16026.

Better Humans describes itself as "an editorial production company that's dedicated to having the best information, analysis and opinion on the impact of advancing science and technology." The website, www.betterhumans.com, carries many valuable stories, whose facts can generally be trusted. Less so, the opinions, but it's easy to tell the difference.

The World Transhumanist Association has a large and well-ordered site, transhumanism.org. Secretary James Hughes has his own website, with challenging essays, including the greatest title in the field, "Embracing Change with All Four Arms: A Post-Humanist Defense of Genetic Engineering," at www.changesurfer.com/Hlth/Genetech.html.

See also www.extropy.org, www.maxmore.com, www.promethea.org, www.prometheism.net, and the *Reason* magazine archive at www.reason.com/biclone.shtml.

Books

Lee Silver, *Remaking Eden: How Genetic Engineering and Cloning Will Transform the American Family*, Bard, 1998; originally published as *Remaking Eden: Cloning and Beyond in a Brave New World*, Avon, 1997

Gregory Stock and John Campbell, eds., *Engineering the Human Germline*, Oxford University Press, 2000

Gregory Stock, *Redesigning Humans: Our Inevitable Genetic Future*, Houghton Mifflin, 2002. The 2003 paperback edition, published by Mariner Books, has a different subtitle: *Choosing Our Genes, Changing Our Future*.

James Hughes, *Citizen Cyborg: Why Democratic Societies Must Respond to the Redesigned Human of the Future*, Westview Press, Cambridge, MA, 2004.

Brian Alexander, *Rapture: How Biotech Became the New Religion*, Basic Books, New York, 2003

Stephen S. Hall, *Merchants of Immortality: Chasing the Dream of Human Life Extension*, Houghton Mifflin Company, New York, 2003

■ ENDNOTES ■

1 David Ewing Duncan, "Reversing Bad Truths," *Discover*, 07/03

2 Jonathan Weiner, *Time, Love, Memory*, Knopf, 1999

3 Ralph Brave, "James Watson Wants to Build a Better Human," AlterNet, 05/28/03

4 Tom Abate, "Nobel Winner's Theories Raise Uproar in Berkeley: Geneticist's Views Strike Many as Racist, Sexist," *San Francisco Chronicle*, 11/13/00

5 Weiner, op. cit.

6 Jonathan Ewbank, letter to *Nature*, 04/16/98

7 Duncan, op. cit.

8 Shaoni Bhattacharya, "Genetically Enhanced Humans to Come, Say DNA Pioneers," *New Scientist*, 04/24/03

9 Gregory Stock and John Campbell, eds., *Engineering the Human Germline: An Exploration of the Science and Ethics of Altering the Genes We Pass to Our Children*, New York: Oxford University Press, 2000

10 ibid.

11 Anne McLaren, letter to *Nature*, 04/16/98

12 Quoted by Brave, op. cit., from James D. Watson and Andrew Berry, *DNA: The Secret of Life*, Knopf, 2003

13 Weiner, op. cit.

14 Walter Gilbert, quoted in Brave, op. cit.

15 Duncan, op. cit.

16 Weiner, op. cit.

17 American Heritage Electronic Dictionary, 3rd edition, 1992

18 Susan Lindee, "Watson's World," *Science*, 04/18/03

19 www.research.arc2.ucla.edu/pmts/Stock.htm

20 David P. Hamilton, "In the Debate on Cloning Humans, UCLA Professor Is One of a Kind," *Wall Street Journal*, 06/13/02

21 www.genetics-and-society.org/analysis/promoencouraging/trans human.html

22 Ronald Bailey, "Making the Future Safe," *Reason*, 07/02/03

23 Daniel Sneider, "Rush by States to Ban Cloning Draws Ire, Again and Again," *Christian Science Monitor*, 03/21/97

24 At a hearing of the California Senate Committee on Health and Human Services, 03/20/02

25 Paul Berg, "Testimony on Nuclear Transplantation before Senate Health, Education, Labor and Pensions Committee on March 5, 2002," available at the American Society for Cell Biology website, www.ascb.org/public policy/bergtest.htm

26 See, for instance, a scathing article by the Cato Institute's Steven Milloy, "Stem Cell Panel Has Vested Interest in Research," Fox News, 01/25/02

27 Robert Langreth and Stephen D. Moore, "Not Proud of It Right Now," *Wall Street Journal*, 10/27/99; available at www.contac.org/contac library/research34.htm

28 Gretchen Vogel, "California Debates Whether to Become Stem Cell Heavyweight," *Science*, 09/10/04

29 From the press release, which includes the script, at the "Yes on 71" website, www.yeson71.com/news_rel_0923.php, checked 11/24/04; it will presumably not last long.

30 Option price from Milloy, op. cit.; stock prices from Andrew Pollack, "Measure Passed, California Weighs Its Future as a Stem Cell Epicenter," *New York Times*, 11/04/04

31 American Heritage Electronic Dictionary

32 Weiner, op. cit.

33 Neil Munro, "Petri-Dish Politics," *National Journal*, 04/19/03

34 Libertarian Party Press Release, "Don't Play God With Human Cloning, Libertarian Party Warns Politicians," 02/25/97

35 Ronald Bailey, "Split Decision: The reckless conservatism of the President's Council on Bioethics," *Reason*, 07/17/02

36 Alex Epstein, "The Virtue of 'Playing God,'" 06/17/02; www.aynrand.org

37 Lee Silver, *Remaking Eden: How Genetic Engineering and Cloning Will Transform the American Family*, Bard, 1998

38 Bradford McKee, "In This Ring, a Designer Slugfest," *New York Times*, 11/04/04

39 Extropians FAQ List (release 0.092), downloaded from www.extropy.org 03/11/00; FAQ rechecked and searched 11/23/04

40 The full essay, "A Letter to Mother Nature," is at www.maxmore.com.

41 ExiCommunity Polls, cited in James Hughes, "The Politics of Transhumanism," a paper originally prepared for the 2001 Annual Meeting of the Social Studies of Science

42 www.homoexcelsior.com/archive/transhuman/msg03942.html

43 Hughes, op. cit.

44 www.transhumanism.org/tv/2004/presenters.shtml and transhumanism.org/tv/2003usa/speakers.htm

45 Ronald Bailey, "The Transhumans Are Coming!" *Reason*, 08/11/04; www.reason.com/rb/rb081104.shtml

46 www.promethea.org/Author/Author.html

47 www.prometheanmovement.org/info/faqs.html#eugenics

48 On the front page of www.prometheism.net

49 Ted Agres, "US Congress Passes FY05 Budgets," *The Scientist*, 11/24/04

50 Data from the Foundation Center, www.fdncenter.org

51 The insults can be found in an article reposted, with typical panache, on the University of Phoenix website: Paul Keegan, "Essay Question: The Web Is Transforming the University. How and Why? (Please Use Examples.)" *Business 2.0*, 12/00; available at www.university-of-phoenix-adult-education.org/university_of_phoenix_articles_business_2.0_the_web_is_transforming_the_university5.html

52 Jim Drinkard, "Independent Voices Rising in Ads," *USA Today*, 08/18/04

53 Brian Alexander, *Rapture: How Biotech Became the New Religion*, Basic Books, New York, 2003, especially pp. 235–43

54 Gersh Kuntzman, "Just Cloning Around," *Newsweek*, 10/12/04. The company website is savingsandclone.com.

55 Interview in *Japan Today*, 11/05/04

56 Brian Alexander, "John Sperling Wants You to Live Forever," *Wired*, 02/04

57 Scott Burns, "Sir John Templeton: Keep a Cautious Watch," *Dallas Morning News*, 11/22/04

58 Financial data from GuideStar (www.guidestar.org) and the Foundation Center

59 Faye Flam, "Suddenly Science, Religion Are Communing with One Another," *Philadelphia Inquirer*, 05/07/99

60 Lawrence M. Krauss, "An Article of Faith: Science and Religion Don't Mix," *Chronicle of Higher Education*, 11/26/99

61 "The World's Richest People," *Forbes*, 02/26/04

62 Jim Doyle, "The Burden of Immortality: Slowing the Aging Process Gives Birth to Ethical, Sociological Questions," *San Francisco Chronicle*, 04/25/04

63 Gregory M. Lamb, "High-Tech Hope: Escape from the Sands of Time," *Christian Science Monitor*, 08/05/04

64 Audrey Gillan, "Pssst . . . the Secret of Youth Can Be Yours for £250," *Guardian*, 09/11/04

65 See www.ucsf.edu/pibs/faculty/kenyon.html, for references.

66 www.elixirpharm.com

67 Matthew Herper, "Cloning People Easier Than Profits," *Forbes*, 12/30/02

68 The company website is www.chromos.com; for a brief discussion of the implications of artificial chromosomes, and links, see www.genetics-and-society.org/analysis/biotech.html

69 "Daughter of Ted Williams Drops Case Against Alcor," AP, 06/15/04

70 Raja Mishra and Beth Daley, "Williams Dispute Piques Interest in Cryonics, Except Among Scientists," *Boston Globe*, 07/11/02

71 Stephen S. Hall, *Merchants of Immortality: Chasing the Dream of Human Life Extension*, New York: Houghton Mifflin Company, 2003, especially pp. 67–68

72 Ed Welles, "Who Is Doctor West? And Why Has He Got George Bush So Ticked Off?" *Fortune*, 04/01/02

73 Hall, op. cit., pp 78–79

74 The $40 million figure is from a lecture West gave on 10/12/04. The $2 million is from an "Entrepreneur Profile" titled "Cloning: Huckster or Hero?" in *Business Week*, 06/20/02.

75 Hall, op. cit., p. 84

76 *Business Week*, op. cit.

77 These were all widely reported but a useful summary is: Porter Anderson, "Company Behind the Clones: Advanced Cell Technology," CNN, 11/25/01; www.archives.cnn.com/2001/TECH/science/11/25/cloning.act/

78 The Jaenisch quote is from Welles, op. cit. The trifecta of scientific, pop-science, and newsmagazine articles is: Jose B. Cibelli, et al., "Somatic Cell Nuclear Transfer in Humans: Pronuclear and Early Embryonic Development," e-biomed: the journal of regenerative medicine, 11/25/01; Jose B. Cibelli et al., "The First Human Cloned Embryo," *Scientific American*, 11/24/01; Joanie Fischer, "The First Clone," *U.S. News & World Report*, 12/03/01.

79 Robert A. Weinberg, "Of Clones and Clowns," *The Atlantic Monthly*, 06/02

80 Tom Abate, "Cloning Firm Says Focus Is on Stem Cells, Not Embryos," *San Francisco Chronicle*, 11/29/01

81 Antonio Regalado, "U.S. Cloning Pioneer Resigns to Accept University Position," *Wall Street Journal*, 11/22/02

82 www.advancedcell.com, checked 11/20/04

83 www.bioethics.upenn.edu/faculty/index.php?profile=1

84 www.washingtonspeakers.com/speakers/speaker.cfm?SpeakerId=3105

85 Arthur Allen, "Bioethics Comes of Age," *Salon*, 09/28/00

86 Jennifer Lee Atkin, "Educators and Experts," *Science & Spirit*, 11–12/01

87 Arthur L Caplan, Glenn McGee, and David Magnus, "What Is Immoral About Eugenics?" *British Medical Journal*, 11/13/99

88 In an interview for ABC, quoted at www.genetics-and-society.org/overview/threshold.html

89 Lori Andrews, *The Clone Age: Adventures in the New World of Reproductive Technology*, New York: Henry Holt and Company, 1999, p. 206

90 Sarah Sexton, "If Cloning Is the Answer, What Was the Question?" Corner House Briefing 16, 10/99; www.thecornerhouse.org.uk/

91 Wicker has several websites, including www.humancloning.org and www.clonerights.com.

11

PRESERVING OUR COMMON HUMANITY

■

INTRODUCTION

SOCIAL JUSTICE

DISABILITY AND DISCRIMINATION

FEMINIST CRITIQUES

ENVIRONMENTALIST CRITIQUES

CONSERVATIVE AND RELIGIOUS CRITIQUES

MOVING FORWARD TOGETHER

FURTHER READING

■

■ INTRODUCTION ■

O PPOSITION TO Human GE springs from a wide variety
of sources. Taken together, what they generally have in
common is a sense of collective responsibility—a communi-
tarian approach to social and political issues, rather than a libertarian, individualistic view.

Most activists, however, develop their views not from any single overarching theory but from one or more specific critiques,
most of which do turn out to be mutually compatible. Several
were considered in specific relation to cloning (see **Chapter 3**),
but are generally applicable to Human GE:

> ◗ **repugnance**, an initial "gut reaction" of opposition
> ◗ **safety**, on several different grounds
> ◗ **eugenics** and what cloning and Human GE might lead to
> ◗ **objectification** and **commodification**
> ◗ personal, familial and social **relationship issues**

Without duplicating that discussion—or the criticisms
implicit and explicit throughout the text—this chapter outlines some of the various approaches to critiquing Human GE,
with particular emphasis on the secular ones. There are, for
some people, strictly religious objections, which are not considered here in any great detail: Those who believe them don't
need the discussion; those who don't, probably won't be convinced by them.

Different activist groups, naturally, tend to work on different
aspects, but a few organizations do have Human GE as their
principal focus, or at least a major one. They are especially

recommended for further information and more connections. Among the most useful are:

- The **Center for Genetics and Society** (CGS)—see **Box 11.1**
  www.genetics-and-society.org
- The **Council for Responsible Genetics** (CRG)— founded in 1983, based in Cambridge, Massachusetts, publisher of *GeneWatch* (every two months, in print form, with many of the articles archived on the website); an early critic of genetic reductionism, discrimination, bioweapons, and modern eugenics
  www.gene-watch.org
- **GeneWatch UK** is an unrelated British organization with similar goals, and an almost identical URL
  www.genewatch.org

Other important individuals and organizations are discussed in the context of specific issue areas.

11.1 THE CENTER FOR GENETICS AND SOCIETY

THE CENTER FOR Genetics and Society (CGS) is an Oakland-based non-profit, working "to encourage responsible uses and effective societal governance of the new human genetic and reproductive technologies." CGS developed in 2001 out of the Exploratory Initiative on the New Human Genetic Technologies, created by Richard Hayes and Marcy Darnovsky, who became Executive Director and Associate Executive Director of the new organization.

Hayes has a PhD from the University of California (UC) at Berkeley, and many years of experience as an organizer, including nearly a decade with the Sierra Club. Darnovsky's doctorate is from UC Santa Cruz; her activist experience covers a very wide range with a recent emphasis on feminist and social justice issues.

(Personal note: I've known the principals for several years and done

11.1 contract work for them, including research for their excellent website. They encouraged the writing of this book, commented on parts of it, and are, of course, responsible for none of it.)

CGS has not primarily focused on grassroots organizing, preferring to work in coalition, often behind the scenes. The e-mail newsletter *Genetic Crossroads* is highly recommended; subscribe at www.genetics-and-society.org/newsletter/index.html.

They must be doing something right: Hayes has been described as the "Arch-BioLuddite" by no less an authority than the General Secretary of the World Transhumanist Association.[1]

■ SOCIAL JUSTICE ■

"We hold these truths to be self-evident, that all Men are created equal . . ."

—Declaration of Independence

EQUALITY COMES FIRST in the Declaration of Independence, before any of the talk of rights, which are based on it.[2] This is a creed, an article of faith, a sacred truth of the United States. It is *self-evident*—Jefferson's first draft was "sacred and undeniable"—and as stunning and mysterious a truth as those to be found in, say, the Sermon on the Mount or the *Tao Te Ching*.[3]

But the principle of equality has been under attack, and economic inequality has been rising for a full generation.[4] Human GE is potentially a driver of inequality. In its fundamental theory, it is about control, domination, and the imposition of particular pre-set values, as C. S. Lewis understood more than sixty years ago (see **Box 11.2**).

11.2 ### C. S. LEWIS AND "THE ABOLITION OF MAN"

ONE OF THE most profound essays against human genetic engineering was written by C. S. Lewis a decade before the structure of

11.2 DNA was discovered. "The Abolition of Man" was published in 1943, during World War II, and its first half-dozen pages (the whole essay is only eighteen pages long) say virtually all that is needed on the subject; the rest is commentary.[5]

Lewis considers the "conquest of Nature" and concludes rapidly that, "what we call Man's power over Nature turns out to be a power exercised by some men over other men with Nature as its instrument." In this, he explicitly includes "the power of earlier generations over later ones." He then discusses a time when: "Man by eugenics, by pre-natal conditioning, and by an education and propaganda based on a perfect applied psychology, has obtained full control over himself. Human nature will be the last part of Nature to surrender to Man. . . . The battle will indeed be won. But who, precisely will have won it?"

Lewis means this question in a rather subtle sense. For it is not merely that the species would, presumably, be new—that "Man" would thus be abolished in a strictly scientific sense—but that the purely rational being creating this new species would have no values, indeed no way of making a value judgment, beyond the transient impulses generated by Nature. "Their extreme rationalism, by 'seeing through' all 'rational' motives, leaves them creatures of wholly irrational behavior. . . . Man's conquest of Nature turns out, in the moment of its consummation, to be Nature's conquest of Man." He insists that his is not an attack on science, and suggests, tentatively, "a new Natural Philosophy" of which he says, "When it spoke of the parts it would remember the whole."

Lewis was (in addition to being the author of the "Narnia" novels) one of the great Christian apologists of the mid-twentieth century and an important academic, a Fellow at Magdalen, Oxford, and then Professor of Mediaeval and Renaissance Literature at Cambridge. He appeals, at root, to intuition, and to what he calls "the only known reality of conscience." It seems a flimsy reed on which to erect a structure of values. But in the end it's all we've got, an innate sense of the rightness of things. Why live? Why choose life itself—my own life, let alone anyone else's—rather than death? Because that is the kind of creature I am, that's all. I eat and sleep and live—and think, and feel, and act, and love—there is no choice in this, and no prisonment either. It is, quite precisely, the air that I breathe.

More immediately, Human GE involves leaving people behind. The threat of developing a genetic aristocracy may be a long way off but is enormously antithetical to any sense of equal rights. This has helped to attract the attention of civil rights groups, especially from communities of color.

Memories of such appalling "research" projects as the Tuskegee syphilis studies should haunt everyone. A group of black men with syphilis were monitored, without treatment, for forty years, beginning in 1932. The original goal—to study the progression of the disease—was questionable, but the worst decisions date to about 1945, when treatment with penicillin became available and was withheld.[6] Nevertheless, the study continued, quietly, until the first newspaper reports, in 1970, began to fuel public outrage. It was halted in 1972, and President Clinton finally apologized in 1997. This is part of the reason that the black community has long been particularly wary of such medical experiments, although eugenics laws were certainly applied to poor whites, too, including the Buck family (see **Chapter 8**).

Anecdotal evidence suggests that black feminists have tended to be quicker to see the potential for abuse in Human GE and related activities than many mainstream white feminists. For example, the California Black Women's Health Project was an important member of the progressive coalition opposing California's 2004 stem cell proposition (see **Chapter 12**), where a number of other women's groups either stayed neutral or supported it. This corresponds with what little race-based public opinion data exists (see **Chapter 9**) and also with the class- and income-based data.

West Harlem Environmental Action (WEACT) presented at Columbia University in February 2002 an early and major effort at "a national conference and community dialogue" on "Human Genetics, Environment, and Communities of Color: Ethical and Social Implications."[7] Speakers included Professor

Troy Duster (an expert on both eugenics and race in America) and Debra Harry (of the **Indigenous Peoples Council on Biocolonialism, IPCB**), as well as representatives from both CGS and CRG.

Similar multidisciplinary and multicultural dialogues have continued, for example with the 2004 *Gender and Justice in the Gene Age* Conference. This was an explicitly feminist meeting about the social implications of new reproductive and genetic technologies. It was described by one of the organizers as "the first U.S. meeting in many years to ground these issues in the values and commitments of feminists who work from a global social justice and human rights perspective," and issues of race and class were high on the agenda; people of color were disproportionately represented. This is quite typical: People approaching these issues from perspectives such as feminism and racism often rapidly find their positions converging.

An earlier and larger conference to make similar connections was the 2001 **World Conference Against Racism (WCAR)** in Durban, South Africa. Nobel laureate (for literature) Nadine Gordimer asked whether genetic engineering was "the new face of racism" and speculated about a future in which the "haves" could afford access to modern technologies, while the "have nots"—"principally dark-skinned people"—could not.[8] UNESCO's Jerome Binde said WCAR needed to address genetic engineering to prevent the creation of "a two-track humanity" of superhumans and "sub-humans." Boston University Health Law Professor George Annas, in a powerful speech, warned that,[9] "Without action on the species level genism will eclipse racism as the most destructive disease on the planet."

Racism is now part of the public debate surrounding Human GE. However there is another group, whose experience with eugenics is absolutely personal, that is perhaps less obvious, and in some senses less visible. That is the disability community.

■ DISABILITY AND DISCRIMINATION ■

THERE IS NO DOUBT that the disability rights perspective on opposing Human GE is challenging for many well-intentioned people. Why *wouldn't* "they" want to be like "us"? Aren't "cures" what it's all about? Don't "they" want to be cured, and to spare others their (presumably) awful existence? Well, why not ask?

It turns out that a significant number of "them" do not look at life this way. Consider, for example, the argument for using PGD to pre-diagnose and discard embryos that might develop into people with diseases. What are usually mentioned as examples are single-gene, debilitating or deadly conditions like Tay-Sachs, cystic fibrosis, or Huntingdon's disease. Most scientists—most people—*assume* that this is a good thing. On the other hand, Dr. Tom Shakespeare pointed out in a public debate,[10] "The world [his opponent] has in mind wouldn't have me in it. Or my two kids, or my dad."

Shakespeare is four feet, five inches tall. He was born with achondroplasia, which is a genetic condition. It doesn't seem to have slowed him down much, either socially or professionally. He has a PhD from Cambridge and is Director of Outreach for the Policy, Ethics and Life Sciences Research Institute (PEALS) in Newcastle, UK. This has given him the base to be an important player in British politics on these issues. Another key quote of his, less confrontational perhaps, is: "Beethoven is a classic case. He was a fifth child, and his siblings were affected by deafness, blindness, syphilis and tuberculosis. In the modern era his mother might have been told to have an abortion and we would have lost some of the most inspiring music ever written."[11]

Moreover, some of the conditions that could be tested for, and thus avoided, may not produce symptoms until late in life. Woody Guthrie suffered from Huntingdon's; his last years were very hard, but before that his life and work were inspirational—should he have been prevented from living? Should Stephen Hawking, the famously wheelchair-bound physicist with the electronic voice synthesizer, who suffers from motor neurone disease? Or Winston Churchill, whose regular bouts with depression might well have shown up in some kind of (future) genetic test? Disability activists are doing the rest of us—who like to believe we have no disabling conditions—a great favor by compelling us to confront the reality of our own prejudices and assumptions.

Another activist who refuses to be invisible is the redoubtable Dr. Gregor Wolbring (see **Box 11.3**). Many more can be found through Disabled Peoples' International (DPI), which produced an excellent position statement in 2000, *Disabled People Speak on the New Genetics*. Like any such well-considered statement from any perspective, it includes material, for example on patents and profit motives, that might be thought extremely relevant to those not part of the immediate constituency. Except that we are *all* part of the immediate constituency: None of us is, or ever will be, perfect.

11.3 | **GREGOR WOLBRING, MAKING PEOPLE UNCOMFORTABLE**

GREGOR WOLBRING is ferociously intelligent, often very funny, and completely committed to the cause of people with what he calls "marginalized characteristics"—the disabled.

In the words of a poem by British performance poet Jim Thomas, posted on Wolbring's own website, he is "an obnoxious, unapologetic, revolutionary" given to shouting about "My People" and "this silent holocaust":[12]

> . . ."My People" he says
> "are defined daily by their defects . . .
> are being redefined as unwanted genes
> on kinky chromosomes, fair game to be
> edited out before birth.
> I am an excuse for abortion.
> I am the argument for euthanasia.
> I am a societal burden, a monster
> or worse . . ."

11.3 (The full poem, which provides perhaps a more rounded picture, can be found at www.bioethicsanddisability.org.)

Wolbring describes himself,[13] "I am a thalidomider and a wheelchair user. I am a biochemist and a bioethicist. I am a scientist, a social Entrepreneur and an activist." He is missing both legs and several fingers. Born in Germany in 1962, he moved to Canada in 1992, where he works as a research scientist and coordinates the International Network on Bioethics and Disability, which he founded. Wolbring appears to have no patience for fools. He also seems quite willing to exploit liberal guilt any time he can, and to claim entitlement as well as equality.

He makes people uncomfortable. He makes people think.

One group of the putatively disabled turns the whole PGD debate on its head. Some Deaf people—many from the culture that identifies itself by using the initial capital letter—are arguing for the right to select an embryo that *will* be deaf like the parents.[14] On a lower-tech level, this has already happened: In 2002, a deaf lesbian couple chose a deaf male friend to be a sperm donor in the hope of developing a deaf baby; the child did indeed turn out to be hearing impaired, "deaf enough."[15] The parents said that if he wants a hearing aid later, they'll give him one; they don't use them. They don't think they're necessary.

To be fair, the disabled community is split on these issues. There are those, like the late Christopher Reeve, who focus on cures.[16] But it is vitally important to pay attention to the point of view that is underrepresented in mainstream society. "They" (the marginalized) see "us" and "we" (the self-defined "normal") need to see "them." This would perhaps be particularly useful for the participants at a crypto-eugenic conference in late 2004, who provoked this comment from British activist Bill Albert,[17] "The one thing I came away with was that these people need to work some changes on themselves so they can grow longer arms so as to better pat themselves on the back for being so right about everything."

A good summary of the disability critique of prenatal selection can be found at the website of Our Bodies Ourselves.[18]

Their approach comes from the women's health perspective; they inevitably end up in the same nexus of issues.

■ FEMINIST CRITIQUES ■

COMMODIFICATION, commercialization, objectification, elitism, and the tyranny of expectations based on appearance do not *have* to be women's issues specifically, any more than wilderness preservation or climate change are. But they *are* issues that have been central to the women's movement for thirty years or more, and feminists have particular expertise in defining and addressing them.

Then there are the facts of women's experience. In the fantasies of Human GE advocates, we are frequently asked to picture a healthy, young, affluent, heterosexual couple leafing through the designer baby catalog (see **Box 5.1**), to select the traits they wish their child to have. That's as far as they generally describe the process, preferring to fast-forward to the perfect baby, if not beyond. But—assuming everything goes to plan—what happens in between? The experience is rather different for the two partners.

For the man:

▶ He will masturbate when called upon.

For the woman:

▶ She will be fed drugs to overstimulate her ovaries.
▶ Her eggs will be removed surgically.
▶ She will undergo a surgical implantation.
▶ She will then carry the fetus to term, with all the usual risks and perhaps a few more.

Of *course* these are women's issues. It's women who are at risk.

Some of this applies to all assisted reproduction technologies (ART). Indeed, Dorothy Roberts has pointed out (in "Race and the New Reproduction," yet another example of the interconnectedness of these analyses) that it may be disproportionately men that benefit from ART, because,[19] "the new reproduction enforces traditional patriarchal roles that privilege men's genetic desires and objectify women's procreative capacity. . . . High-tech procedures resolve the male anxiety over ascertaining paternity: by uniting the egg and sperm outside the uterus, they '[allow] men for the first time in history, to be absolutely certain that they are the genetic fathers of their future children.'"

Still, infertile couples may well consider the risks—and costs—to be worth taking. Human GE, however, raises the stakes, just on a practical level. The likelihood of problems is certainly greater, as is the likely requirement for women's eggs, if they are to be experimented upon (see **Chapter 2**). The concerns about objectification and the rest increase too, as does the issue of consent (see **Chapter 3**).

Some abortion rights groups have chosen to stay out of the debates over cloning and Human GE, for fear that any regulation will impinge on hard-won—and threatened—reproductive rights. However, Marcy Darnovsky has noted that, "Abortion rights groups will . . . increasingly be drawn into the politics of genetic manipulation because supporters of germline engineering and cloning have taken up their language, keywords, and appeals."[20]

For example, one quartet of bioethicists—all men—sympathetic to Human GE titled a book *From Chance to Choice*, in a blatant attempt to appeal to abortion rights supporters.[21] More generally, enthusiasts routinely try to claim both cloning and GE as "reproductive rights." This has been denounced by many feminist leaders, over 100 of whom signed a statement in 2001 particularly noting that,[22] ". . . cloning advocates are seeking to appropriate the language of reproductive rights and freedom of choice to support their case. This is a travesty, and needs to be challenged. There is an immense difference between ending an unwanted pregnancy and creating a duplicate human. Most people readily understand this, and can support abortion rights while opposing human cloning."

Those advocates are almost all men; the whole biotechnology industry is dominated by men, with one study finding only 20 out of 265 directors (7.5 percent) to be women, none of them "chairmen," and only 40 of 263 corporate officers (15 percent), mostly vice presidents.[23] Women tended to work, if at all, in "communications, human resources, facilities, regulatory affairs and scientific departments" but not management.

There does exist a range of opinion among feminists, as Lisa Handwerker has noted, "influenced in part, by our multiple identities including whether or not we are: rich, poor, rural, urban, lesbian, transgender, heterosexual, disabled, Caucasian, Jewish, women of color, multi-ethnic, religious or non-religious, pre- or post-menopausal, young, mid-life, old, fertile and/or infertile."[24] There are even "cyborg feminists" who are actively interested in becoming part-machine/part-human. (If this only affects the individual and not her offspring, it would not be incompatible with opposition to germline manipulation.)

Handwerker argues that "while respecting these differences, . . . we need to consider a united position, which opposes human reproductive cloning and germ line alteration." One significant reason she cites is that, "Scientists and doctors in support of human reproductive cloning, I believe, inadvertently feed into anti-abortion politics by further delineating the separation of a fetus from the woman." (Cloning would likely *increase* the need for late-term abortions, if animal experience is any guide, since cloned animal embryos sometimes begin developing normally before things go awry.)

The entire focus of the genetic modification effort works to place the attention away from the woman who is or may be pregnant and onto the unborn and putatively perfect child. Motherhood thus reverts from the liberated and liberating natural process that has been a goal of the women's movement, back to the old patriarchical experience of marital relations, not to mention "those doctors who were condescending, paternalistic, judgmental and noninformative."[25]

Which is exactly what brought the **Boston Women's Health Book Collective (BWHBC)** together in the late 1960s. The long-standing women's health group became famous for *Our Bodies, Ourselves* (OBOS, often used to name the collective too; see www.ourbodiesourselves.org), which was first published in 1970; the latest English edition came out in 2005, following on at least nineteen foreign translations. OBOS is firmly against the abuse of human genetic technologies; cofounder and Executive Director Judy Norsigian has been a prominent speaker against reproductive cloning and unnecessary and unregulated experimental procedures.

Another organization working in this area is **The Committee on Women, Population, and the Environment (CWPE)**. This multiracial alliance of feminist activists, health practitioners, and scholars has been actively involved in the critique of the use of PGD for sex selection (see **Chapter 5**). Among their prime concerns are the environment, violence, immigration, and a raft of issues related perhaps by manifestations of racism, anger, and hate. And they closely monitor genetic and reproductive technologies. For more information, and background materials, see www.cwpe.org.

Cloning, in particular, is "not a women's issue alone, but that is a piece of it," said Norsigian in one interview.[26] "The big question is," said Lori Andrews once, "what kind of society do we want to be in?"[27] The feminist approach is one important road toward an answer.

■ ENVIRONMENTALIST CRITIQUES ■

ANOTHER MAJOR APPROACH to the issue is **environmentalist**. Ecological awareness and familiarity with the precautionary principle are enough by themselves to make some people very reluctant to proceed with GE of any kind. Environmentalists were among the first people to express concerns, back in the early 1970s, possibly even before Jeremy Rifkin (see **Box 11.4**) began his long-running series of campaigns. Certainly the GE food campaigners (see **Box 11.5**) are supportive of efforts to curb Human GE.

THE INDEFATIGABLE JEREMY RIFKIN

JEREMY RIFKIN has been working to stop the abuse of biotechnology for close to thirty years. He long ago earned such sobriquets as:[28]

- "The most hated man in science" (Time magazine)
- "biological fundamentalist" (Nobel laureate David Baltimore)
- "food terrorist" (head of the National Milk Producers Federation)

Rifkin made his bones as an activist with the 1968 March on the Pentagon, and then the People's Bicentennial Commission.[29] In 1977 he established the Foundation on Economic Trends (FoET) in Washington, D.C., which serves as his organizational base (though he is often regarded—like Ralph Nader—as a solo operator). That same year, he and others interrupted a National Academy of Science meeting with chants of, "We shall not be cloned." This is thought to be the first anti-biotech demo. It unnerved some of the scientists present, who considered themselves to be "hostages" and "terrorized," though all that happened was a speech.[30]

Theological Alliances

Rifkin was an early and persistent builder of coalitions with religious groups. In 1982, he wrote a 10-page "Theological Letter Concerning the Moral Arguments Against Genetic Engineering of the Human Germline Cells" and sent that to religious groups. This said in part,[31] "Genetic engineering of the human germline represents a fundamental threat to the preservation of the human species as we know it, and should be opposed

11.4

with the same courage and conviction as we now oppose the threat of nuclear extinction." Rifkin's organizing led to a resolution against germline manipulation, which was signed by what the New York Times, in a front page article, called an "unusually broad spectrum of religious leaders and several prominent scientists."[32] It was presented to Congress on June 8, 1983.

A 1987 coalition opposed the patenting of genetically modified animals.[33] So too did a 1995 effort, which originated with the United Methodist Church. That one was signed by "roughly 100 Roman Catholic bishops, numerous Protestant and Jewish leaders and groups of American Muslims, Hindus and Buddhists."[34]

The Unstoppable Activist

Rifkin himself continues to write—The Biotech Century is the most specifically relevant of his many books—campaign and stir up controversy. Critics sometimes object that he has been saying the same thing for years—but if he was right then and is right now, why on earth should he change his mind? Those who try to discredit him generally haven't.

The magnificently appalling website www.activistcash.com (a hypocritical front for various industry groups) can provide one final, gloriously gratuitous, badge of honor: "Next to the Unabomber, Rifkin is perhaps America's most notable anti-technologist."

11.5

THE FOOD CAMPAIGNS

ALTHOUGH HUMAN GE was an early focus of activism, agricultural biotech became a more immediate problem. In response, activists developed a number of organizations to confront GE Food (see the Appendix), supplementing efforts by some previously established organizations such as Greenpeace and Friends of the Earth (FoE).

All the GE activists are concerned about human as well as agricultural applications. For example, the Convergence in San Francisco in June 2004, a large gathering organized in opposition to the annual biotech industry conference, was titled "Reclaim the Commons" and publicity featured this litany,[35] "Engineered Foods? Medicine For Profit? Corporate Control? Designer Babies? Biological Weapons?"

11.5 One of the most encouraging aspects of the food campaign is the insistence of organizers on linking agribusiness to issues of war, trade, globalization, and social justice. The Sacramento Mobilization the year before, which was specifically focused on food, since it countered an Agriculture Department meeting focused at selling GE products abroad, also included workshops on Human GE, as well as many other, issues.[36]

There is a populist, grassroots element to the GE Food campaign that has been—so far—less apparent in most events specifically focused on Human GE. Conversely, the Human GE campaign may have more sympathetic connections deep in the Bush administration. The campaigns can only benefit each other.

Other prominent environmentalists who have taken public stands against germline engineering include Amory Lovins, Terry Tempest Williams, Gary Snyder, and Mark Dowie, who were among close to 250 notables who signed an Open Letter on the subject in February 2000, calling for an outright ban.[37] Bill McKibben, who made his name with *The End of Nature*, an excellent book mostly about global warming, turned his attention to Human GE with the fine *Enough: Staying Human in an Engineered Age*.[38]

Organizationally, groups such as Friends of the Earth (FoE), Physicians for Social Responsibility (PSR), Greenpeace, and the Sierra Club have become involved to varying degrees. Brent Blackwelder, the President of FoE, and Robert Musil, the Executive Director of PSR, sent a long letter to Congress, the President, and federal agencies, in which they said, in part, "While all of us seek to improve the quality of human life, we believe that certain activities in the area of genetics and cloning should be prohibited because they violate basic environmental and ethical principles—principles which form the core values for which our organizations stand."[39]

The first mentioned of those is the precautionary principle, the environmental equivalent of the old doctor's adage, first do no harm.[40] The basic requirement of this is that we know what will happen before we do it—or at least know the extent of the risks of harm so that they can be properly evaluated.

"That does not mean 'do not take any risks at all,'" points out Tewolde Berhan Gebre Egziabher, the Ethiopian expert on biodiversity and related issues, ". . . It means, rather, that if you must take a risk in the absence of knowledge you should err on the side of caution and say, until I know, I won't do this.'"[41] Human GE is a long way from demonstrating enough knowledge to validate going ahead, and unlikely to get close unless we attempt unpredictable, hence unethical, experiments on people. In the case of a terminally ill patient, the risks involved might be acceptable; in the case of an unconceived child they cannot possibly be.

When biotechnologists alter plants and animals, they claim to do it for people. No one pretends that a pest resistant crop is a happy crop, or that a cloned cow would be a contented cow. Their whole discussion is external, is about one species doing something to another. To them, it is as though the environment is "out there," and consists of everything except ourselves; wilderness, a nice place to visit.

But Human GE cuts through all that. This is our own species that we are proposing to alter. How can we stand outside and analyze what we inescapably are? We do it all the time, of course—those are the compromises we make whenever we consider the outside world—but this is emotionally difficult terrain. We find it hard to deal with this issue.

This does go to the absolute core of environmental principles. We are one species, with one genetic heritage, a commons we share. Any individual may fiddle with their own individual expression of it, be it by deliberate bodily mutation or adornment, or even (on medical grounds) by having their own genetic structure altered. But when we bid to alter the commons, we cross a line that environmentalists should readily recognize.

The baby, the student, the bride, the mother, and grandmother, these are the same person, and we understand this deeply. It is profoundly disorienting to visit someone suffering from the early stages of Alzheimer's and to discover that they know who you are, but not *when* you are; your identity but not your circumstances. It is almost as surprising to see, for the first time in decades, a childhood friend grown old—and to know immediately who that person is. They've changed, but they're still the same.

There's a kind of comfort, too, that comes from walking in the ruins of the Acropolis or Angkor Wat or Stonehenge or Machu Picchu and feeling that "we were here"—not perhaps my specific ancestors, but ours, the people of old. That is the connection that germline engineering aims to break, to cut as casually as we might snap a piece of string.

Every standard environmental argument applies to the issue of human genetic engineering. Species preservation, biodiversity, heritage, the wonder and beauty of nature, all call for the drawing of a firm line before anyone is allowed to change who everyone is. "No man is an island," wrote John Donne, and he was right. We have a common genetic heritage, our gene pool is a genetic commons, and no individual has the right to pollute it. "Save the Humans!" was a common joke back in the 1970s. It has become all too serious now.

■ CONSERVATIVE AND RELIGIOUS CRITIQUES ■

THE RESPECT FOR the natural world—a sense of the fitness of things—that is clearly violated by many applications of Human GE is also shared by conservatives of many stripes. It is *not* shared by the economic radicals who seek to abolish the social gains of the last seventy years, nor by "free market" capitalists who wish to exploit the environment. It *is*, however, part of the long-standing critique made by the likes of Leon Kass (see **Box 11.6**), whose status as an articulate advisor to President Bush has made him the focus of extraordinarily intemperate criticism.

11.6 LEON KASS, LIGHTNING ROD

LEON KASS is a medical doctor by first training, who then took a PhD in biochemistry at Harvard and only then shifted his interest to philosophy and bioethics, eventually becoming Addie Clark Harding Professor in the Committee of Social Thought at the University of Chicago. He is also a fellow of the American Enterprise Institute, and a prolific author. Among many others, he wrote the remarkable essay "The Wisdom of Repugnance" (see **Chapter 3**); his books include *Toward a More Natural Science: Biology and Human Affairs*.

Kass has been perhaps the leading conservative theorist on Human GE and related matters for years; when President Bush appointed him Chairman of the President's Council on Bioethics in 2001, he could effectively claim the title officially. Certainly, his critics like to say so, and regularly accuse him of packing the Council to reflect his own views. Kass, however, strongly disagrees, seeing his role as,[42] "not to see that my view prevails. My job here is to see to it that we provide the president with the richest possible consideration, so that he knows what is at stake in whatever decision he makes."

True religious conservatives, it seems, are not all that sure about the soundness of his views. This is the Catholic Richard Doerflinger, who knows him well and respects him greatly:[43] "If pushed to the wall, he would say [that an early-stage embryo is] a human organism that has some claim on our respect, but he would not say this is a full person."

Indeed, in 2004, Kass came under fire when his committee recommended (among other proposed regulations) that Congress should prohibit the use of human embryos in research beyond a designated stage in their development (between 10 and 14 days after fertilization).[44] The problem? That implies that scientists may use embryos in research up to that stage. Rather than seeing this as an achievable compromise, some social conservatives saw it as a betrayal.[45]

Kass, who is incidentally not a fundamentalist Christian (he is Jewish), has views on marriage, the family, and homosexuality that are outrageously anachronistic, and he knows it. Surely he chuckled as he ended an essay on "The End of Courtship" with this:[46] "Is there perhaps some nascent young feminist out there who would like to make her name great

11.6 and who will seize the golden opportunity for advancing the truest interest of women (and men and children) by raising (again) the radical banner, 'Not until you marry me'? And, while I'm dreaming, why not also, 'Not without my parents' blessings'?"

He drives *everyone* nuts. And with all that, he seems to be a gentleman of the old school. Kass may not entirely deserve his role as lightning rod, but he wears it well.

Many conservatives base their objections to Human GE on strictly religious grounds. There are a few religious theorists who consider that humans have the right—or even duty—to remake ourselves genetically. The overwhelming majority, however, are at least extremely skeptical if not actively opposed to germline manipulation. Some are willing to accept research cloning (for medical reasons) but very few will countenance reproductive cloning. Exceptions include some Jewish theologians and also some Muslims, although they temper their opinions with social justice concerns:[47] "In view of limited resources in the Islamic world and the expensive technology that is needed for research related to cloning, Muslim legists have asked their governments to ban research on cloning at this time."

The Conference of Catholic Bishops, United Methodist Church, Southern Baptist Convention, United Church of Christ, Episcopal Church, and other Christian denominations have all developed positions on these issues. Their websites are listed in the Appendix; CGS has a summary, under Perspectives/Religious Communities (www.genetics-and -society.org/perspectives/religious.html). The same page includes links to Jewish, Islamic, Hindu, and ecumenical sites.

Many religious conservatives are adamantly opposed to abortion, while virtually all progressives support freedom of choice. There are those, however, attempting actively to get past this disagreement. Any talk of alliance of pro-choice and anti-abortion activists on an issue as sensitive as reproduction might be expected to raise eyebrows, and it has (see **Box**

11.7). Nevertheless, such discussions continue, and it's time to get past the expressions of surprise.

NATURAL ALLIES

ADVOCATES OF Human GE frequently overstress the religious objections—not exactly as a "straw figure" since they are real, but as a favorite target—to the exclusion of secular, progressive ones. Their argument is largely directed at the left, and says, in effect: If religious conservatives are against Human GE, then surely all secular liberals must be in favor. This is pernicious nonsense.

The focus on religious conservatives was never enough to fool the media into thinking there was no other opposition. In particular, feminists and other progressives raised objections to, and called for a moratorium on, embryonic stem cell research in 2001–2. This led to a rash of newspaper and magazine articles, in the San Francisco Chronicle, New York Times, and Wired among many others, talking about "strange bedfellows."[48]

Not only does this cliché deserve to be sent back to Shakespeare's Tempest where it belongs (Act II, scene 2), it really shouldn't be considered relevant. Why on earth should secular progressives and devoutly religious people not work together for the betterment of society? The religious have always been a vital part of progressive movements.

Admittedly, the religious opposition to Human GE, which is well-organized and vocal, tends to come from the more conservative denominations, while the religious involvement in more conventionally progressive causes tends to come from the more liberal—but these floating, informal alliances are never as simple as their opponents would like.

What the GE advocates have tried to do, with some success, is frame the discussion around abortion rights and the freedom to choose. That works for them, by and large. But most people have no trouble distinguishing between ending an unwanted pregnancy and manufacturing a specified child. Preserving our common humanity is a goal we can all share.

■ MOVING FORWARD TOGETHER ■

ONE INSTITUTIONAL INITATIVE aimed at bringing together people from different perspectives to discuss regulations is the **Institute on Biotechnology and the Human Future** (IBHF; www.thehumanfuture.org). IBHF is affiliated with the Illinois Institute of Technology, which includes the Chicago-Kent College of Law, the home base of the staunchly feminist Professor Lori Andrews. She cofounded the IBHF with Nigel Cameron, a Christian bioethicist and Dean of the Wilberforce Forum.

Cameron and Andrews caused something of a stir when they cowrote an article for the *Chicago Tribune* in 2001. They did use the "strange bedfellows" approach, listing them in some detail— Judy Norsigian of OBOS; Richard Doerflinger of the National Conference of Catholic Bishops; Francis Fukuyama, the conservative theorist; Stuart Newman, the pro-choice, anti-cloning scientist from CRG. They concluded, optimistically:[49]

> But when history is written, we harbor no doubts that the cloning debate of 2001 will be noted as the start of something very big, in which those who oppose abortion and those favor reproductive rights discovered common ground in their commitment to the human future and the distrust of uncontrolled biotechnology, and revealed the extraordinary potential of their working together.

■ FURTHER READING ■

Free Documents from the Web

Dorothy Roberts, "Race and the New Reproduction," Chapter 6 of *Killing the Black Body: Race, Reproduction and the Meaning of Liberty*, New York: Pantheon, 1997; available at www.hsph.harvard.edu/rt21/race/ROBERTS6.html

Brent Blackwelder, "Cloning, Germline Engineering, Designer Babies, and The Human Future," Remarks at 50th Anniversary of the Law-Medicine Center, Case Western Reserve University, Cleveland, 10/08/03, available at www.thehumanfuture.org/commentaries/blackwelder_humanfuture.htm

Gregor Wolbring's website, www.bioethicsanddisability.org, includes articles by him and extensive links, by no means all to those who agree with him.

George J. Annas, "Genism, Racism, and the Prospect of Genetic Genocide," prepared for presentation at UNESCO 21st Century Talks: The New Aspects of Racism in the Age of Globalization and the Gene Revolution at the World Conference against Racism, Racial Discrimination, Xenophobia and Related Intolerance, Durban, South Africa, 09/03/01; available at www.thehumanfuture.org/commentaries/annas_genism.htm

John H. Evans, *Cloning Adam's Rib: A Primer on Religious Responses to Cloning*, The Pew Forum on Religion and Public Life, 2002; available at www.pewtrusts.com/pdf/rel_pew_forum_adams_rib.pdf

Books

See also those listed elsewhere, especially in Chapters 1 and 2 and the Appendix.

C.S. Lewis, *The Abolition of Man*, first published in Great Britain in 1943 by Oxford University Press, since reprinted many times

Ruth Hubbard, *Profitable Promises: Essays on Women, Science and Health*, Common Courage Press, 1995

John H. Evans, *Playing God? Human Genetic Engineering and the Rationalization of Public Bioethical Discourse*, University of Chicago Press, 2002

Richard Heinberg, *Cloning the Buddha: The Moral Impact of Biotechnology*, Quest Books, 1999; a genuinely different book that largely succeeds in balancing factual reporting with a perspective that occasionally threatens to drift away into Gaian whimsy but remains rooted in American politics.

Linda K. Bevington et al., *Basic Questions on Genetics, Stem Cell Research, and Cloning: Are These Technologies Okay to Use?* Kregel Publications, 2004; a short paperback in question-and-answer format, from a specifically Christian viewpoint, available from the Center for Bioethics and Human Dignity (CBHD) at its website, www.cbhd.org.

▪ ENDNOTES ▪

1 James Hughes, "Arch Bio-Luddite Richard Hayes Defines CybDem Mission," at www.cyborgdemocracy.net/

2 Jefferson's original made the primacy of equality even clearer, by placing a semicolon before the introduction of rights. It also did not mention a "creator" but said, "from that equal creation they derive in rights inherent and unalienable . . ." See www.leftjustified.org/leftjust/lib/sc/ht/decl/d-draft.html.

3 For the Sermon on the Mount (*The Gospel According to Matthew*, chapters 5–7), the "King James" translation is classic. There are many versions of Lao Tse's *Tao Te Ching* but an excellent—and beautiful—one is by Gia-Fu Feng and Jane English, Vintage Books, New York, 1972.

4 See www.inequality.org.

5 C. S. Lewis, *The Abolition of Man*, Oxford University Press, 1943

6 For an overview, see the official Center for Disease Control (CDC) website at www.cdc.gov/nchstp/od/tuskegee/index.html. Another useful list of resources is at www.gpc.edu/~shale/humanities/composition/assignments/experiment/tuskegee.html.

7 See www.weact.org/genetics/index.html

8 *Genetics Crossroads*, 10/03/01

9 George J. Annas, "Genism, Racism, and the Prospect of Genetic Genocide," 09/03/01

10 The debate, on "The Search for Perfection," was organized by Greenpeace and the *New Scientist* and held in London on 04/30/02

11 "This Brave New World into Which I Wouldn't Have Been Born," *Northern Echo*, 06/30/00

12 Jim Thomas, "Gregor's Poem," available at www.bioethicsanddisability.org/poem.html or www.jimsnail.blogspot.com

13 www.bioethicsanddisability.org/aboutme.html

14 Carina Dennis, "Deaf by Design," *Nature*, 10/21/04

15 Liza Mundy, "A World of Their Own," *Washington Post*, 03/31/02

16 See www.christopherreeve.org/

17 Bill Albert, at www.dpi.org

18 www.ourbodiesourselves.org/gendis.htm

19 Dorothy Roberts, "Race and the New Reproduction," Chapter 6 of *Killing the Black Body: Race, Reproduction and the Meaning of Liberty*, Pantheon, New York, 1997

20 Marcy Darnovsky, "Human Germline Manipulation and Cloning as Women's Issues" in *Sex, Race and Surveillance: Feminist Perspectives from the US*, edited by Jael Silliman and Anannya Bhattacharjee, South End Press, 2001

21 Allen Buchanan, Dan W. Brock, Norman Daniels, and Daniel Wikler, *From Chance to Choice: Genetics and Justice*, Cambridge University Press, New York, 2000

22 See ourbodiesourselves.org/clone3.htm.

23 H. Stewart Parker, "Women in Biotech," *HMS Beagle*, 12/21/01

24 Lisa Handwerker, PhD, M.P.H., "The Implications of Human Reproductive Cloning and Germ Line Alteration for Women and Women's Health: Ten Mis-Conceptions," 02/03/01

25 From the Preface to the 1973 edition of *Our Bodies, Ourselves*, available at www.ourbodiesourselves.org/1973pre.htm

26 Richard Willing, "Odd Mix of Activists Stands Together against Cloning," *USA Today*, 07/16/01

27 Dawn MacKeen, "Cloning Conundrums," *Salon*, 05/03/99

28 Gary Stix, "Dark Prophet of Biogenetics," *Scientific American*, 08/97

29 David A. Ridenour, "Jeremy Rifkin: A 19th-century Luddite Disguised as a 20th-century Moralist," Knight-Ridder Tribune Washington Bureau (DC), 05/08/98

30 Eliot Marshall, "The Prophet Jeremy: Thou Shalt Not Splice Genes," *The New Republic*, 12/10/84

31 The original "Theological Letter" has never been officially published, and the full text is buried somewhere in deep storage, according to an email from FoET Director of Operations Alexia Robinson, 02/11/04.

32 Kenneth A. Briggs, "Clerics Urge U.S. Curn on Gene Engineering," *New York Times*, 06/09/83

33 Richard Stone, "Religious Leaders Oppose Patenting Genes and Animals," *Science*, 05/26/95

34 Edmund L. Andrews, "Religious Leaders Prepare to Fight Patents on Genes," *New York Times*, 05/16/95

35 See www.biodev.org.

36 See, for example, Brian Tokar and Doyle Canning, "Countering Biotech and 'Free Trade' in Sacramento," *Z Magazine*, 09/03

37 Richard Hayes, "The Quiet Campaign for Genetically Engineered Humans," *Earth Island Journal*, Spring 01

38 Bill McKibben, *Enough: Staying Human in an Engineered Age*, Times Books, New York, 2003

39 No date available for this letter; copies were distributed in 09/01 at the "Beyond Cloning" conference in Boston

40 For a full discussion, see "The Precautionary Principle," *Rachel's Environment & Health Weekly* #586, 02/19/98

41 Interview in *Sierra*, 07–08/01

42 Dana Wilkie, "The Puzzle of Leon Kass," *Crisis*, 06/02

43 ibid.

44 The President's Council on Bioethics, *Reproduction and Responsibility: The Regulation of New Biotechnologies*, Washington, D.C., 2004, p. 225

45 Ramesh Ponnuru, "The Kass Council's Ex-Friends," Tech Central Station, 04/20/04

46 Leon Kass, "The End of Courtship," *The Public Interest*, 01/15/97

47 Abdulaziz Sachedina, "Human Clones: An Islamic View," from Glenn McGee, ed., *The Human Cloning Debate*, Berkeley Hills Books, 1998; cited in John H. Evans, *Cloning Adam's Rib: A Primer on Religious Responses to Cloning*, The Pew Forum on Religion and Public Life, 2002

48 A Google search for "strange bedfellows" and "stem cells" on 11/28/04 turned up 631 hits, with little if any duplication.

49 Nigel Cameron and Lori Andrews, "Cloning and the Debate on Abortion," *Chicago Tribune*, 08/08/01

12

REGULATION? WHAT REGULATION?

■

■

■ INTRODUCTION ■

T HE US DOES have *some* regulations about Human GE. But they are deeply confusing, incoherent, and filled with such enormous loopholes that the Food and Drug Administration (FDA), desperate to exert some kind of control, has even been reduced to pretending that cloning is a drug (see **Box 12.1**).

The system is a mess. In fact, system is the wrong word: The US not only lacks a comprehensive oversight mechanism, it lacks any effective process for developing the rules needed to handle the continually changing issues around these technologies. As law professors Lori Andrews and Nanette Elster put it, with academic understatement,[1] "The United States notably lacks an adequate structural mechanism for assessing genetic and reproductive technologies." That's why, as the Center for Genetics and Society has noted, in the absence of either adequate laws or strong regulations, we have seen that:[2]

▶ In 1999 the University of California at San Francisco began secret experiments to clone human embryos, and did not acknowledge this until 2002 when reporters using the California Public Information Act forced disclosure of lab reports.
▶ In 2000 Advanced Cell Technologies in Massachusetts began creating clonal embryos under wraps of corporate secrecy, with no prior review except by an internal "ethics advisory board" known in advance to support research cloning.

12.1 IS CLONING REALLY A DRUG?

In March 2001, the Food and Drug Administration (FDA) took preventive action to stop human reproductive cloning. The agency wrote letters to Dr. Panos Zavos (see **Box 3.8**) and to Dr. Brigitte Boisselier (see **Box 3.6**), warning both of them that "anyone trying cloning must apply for agency permission."[3]

Despite this, Boisselier—head of the Raelian Clonaid enterprise—continued working in a former high school science lab in Nitro, West Virginia, financed by Mark Hunt, a Charleston lawyer.[4]

The FDA leaned on Hunt, who eventually backed out, calling Boisselier "a press hog," and insisting that he had no intention of breaking the law. Richard Seed (see **Box 3.9**) had had a similar experience a few years earlier: "I think their purpose was to frighten me," he later said, "and they did."[5]

The agency was certainly throwing its weight around, and almost everyone would agree it was doing the right thing in acting ahead of time. As US Representative James Greenwood pointed out, an implanted, developing human clonal pregnancy "would pose a fairly difficult enforcement situation."[6] But what legal justification did it have? According to the Boston Globe:[7] "It draws its authority from two laws—the Public Health Service Act and the Food, Drug and Cosmetic Act—that give it jurisdiction over biological products, drugs, and devices. It interprets those laws to cover human cloning both as a biological product and a drug, and insists that it has to give permission to scientists before any research can begin." Cloning is a drug? Or a biological product? Either way is stretching it.

And when it comes right down to it, the FDA knew this perfectly well. What they actually got from Boisselier was an agreement "not to attempt human cloning in the United States, or to do research using human eggs in the United States until the legality of human cloning is ascertained through legislation or a federal declaratory judgment."[8]

It's not even clear they could enforce that, if she really wanted to contest it. But, faced with the options of either fighting the FDA (if she did try to clone) or facing possible fraud charges (if she took money for it but didn't), Boisselier moved Clonaid abroad. Their subsequent—widely disbelieved—announcements always insisted, probably accurately, that no laws had been broken.

▶ In 2001 the Jones Institute for Reproductive Medicine in Virginia broke a widely-observed ethical restraint and harvested 162 eggs from twelve women to create forty viable embryos to be used exclusively for research purposes, again without going through any process of public review and approval.

Each of those activities is problematic in itself, but the more general issue is that there is no standard forum in which to debate them. That means that reputable scientists cannot be confident that their research will be supported, and rogues can blithely ignore anyone else's opinion.

These are issues with which every developed country is grappling. Many countries have passed laws about specific issues such as cloning (see below); others have signed multinational agreements, notably that of the Council of Europe; and a few, including Britain, Canada, Germany, and Australia, have established comprehensive policies and systems from which the US might learn.

■ THE STALEMATE IN WASHINGTON ■

AT LEAST FORTY-FOUR bills to ban cloning were introduced in Congress between 1997 and 2004.[9] Two passed the House, in 2001 and 2003, but none has made it through the Senate.

Both the successful House bills would have prohibited both reproductive cloning and cloning for research. They were proposed by Florida Republican David Weldon, and supported by the party leadership, but did gain votes from almost one third of House Democrats. The 2001 version passed 265–162, the 2003 one 241–155.

The Senate faced competing bills in 1997–8, and again in

2001 and 2003. Each time, the one backed by the Republican leadership called for banning research as well as reproductive cloning, while the one backed by the Democratic leadership only banned reproductive cloning and explicitly or implicitly encouraged the use of cloning techniques for research. Some senators did cross party lines, but essentially (as noted in **Chapter 3**) the biotechnology industry lobbied hard *against* bans on research cloning, and put little or no effort into breaking the deadlock on reproductive cloning.

None of these bills addressed germline engineering, or preimplantation genetic diagnosis (PGD), or any of the other contentious and currently unregulated (or *under-regulated*) issues connected with Human GE. This was, relatively speaking, the easy one. Everyone claimed to be against cloning (if only on safety grounds), but nobody seemed willing to budge.

There were attempts in early 2002 to build support for a compromise that would have banned reproductive cloning and placed a moratorium on research cloning.[10] But those against all embryo research (see **Chapter 4**) thought a moratorium too weak, while, according to Michael Werner of the Biotechnology Industry Organization (BIO),[11] "A moratorium on research is a ban on research, and that is not a compromise to us." The net result is that neither form of cloning is illegal under federal law, and individual states have begun to make their own— contradictory—laws.

▪ STATE LAWS ▪

SEVERAL STATES HAVE passed their own laws about cloning, at least. Those that have addressed the issue have all banned reproductive cloning, but they have taken contradictory positions on cloning for research (see **Table 12.1**).

TABLE 12.1

SUMMARY OF U.S. STATE LAWS ON CLONING[12]

STATE	REPRODUCTIVE CLONING	RESEARCH CLONING
Arkansas	banned	banned
California	banned	allowed
Iowa	banned	banned
Louisiana	ban expired[a]	ban proposed[b]
Michigan	banned	banned
Missouri	use of state funds prohibited	ban proposed[b]
New Jersey	banned	allowed
North Dakota	banned	banned
Rhode Island	banned[c]	allowed
South Dakota	banned	banned
Virginia	banned	unclear[d]

NOTES:

a   A Louisiana law banning reproductive cloning expired in 2003.

b   Louisiana and Missouri are expected to consider proposals to ban or limit research cloning in 2005, as is Kansas.[13]

c   The Rhode Island law will expire on July 7, 2010.

d   The Virginia law may prohibit research cloning, but some experts consider the language imprecise.

California is the state that has taken the lead on this issue. In 1998, the California legislature enacted a five-year moratorium on reproductive human cloning, and established an advisory committee, which was required to report by December 31, 2001 with recommendations on policies to be adopted after 2002. The committee duly reported, recommending a ban on reproductive cloning and "reasonable" regulation of cloning for research. Two laws were passed in 2002, intended to do just that.

The biotechnology industry, through its trade organization, BIO, was strongly behind the California legislation, especially the part about encouraging stem cell research, including cloning. New Jersey passed an almost identical law, and at least five other states—Illinois, Maryland, Massachusetts, New

York, and Washington—were in 2004 considering bills that are word-for-word identical.[14]

But that wasn't enough for some advocates. In 2004, they raised enough money to gather signatures and put on the California ballot Proposition 71, which:[15]

 establishes a constitutional right to stem cell research
 explicitly authorizes cloning research
 sets up a "California Institute for Regenerative Medicine"
 borrows $3 billion to fund it for ten years

Proponents of the initiative outspent opponents by well over 50 to 1 ($34.8 million to $624,973) and it passed handily.[16] In part, this was because Californians took it as a vote against the restrictive policies of President Bush, who lost the state by a substantial margin.

The progressive, pro-choice opposition raised many questions of governance, but they were largely drowned out before the election. Within a month of its passage, however, both the *Los Angeles Times* and *San Francisco Chronicle*—which had endorsed it—were echoing these criticisms. For example:

 There was no guarantee that the state would benefit financially from any joint ventures based on research the people of California had financed;
 nor that national ethical standards about experiments would be maintained (it explicitly allows deviations);
 nor that there would be genuinely independent oversight (the committee charged with that is essentially composed of interested parties, such as scientists and patients' advocates).

State Senator Deborah Ortiz, who had introduced the 2002 cloning legislation and had been a strong supporter of the initiative, suddenly began searching for ways to amend it.[17] This was to prove difficult, since another controversial provision prohibits changes for three years and then requires 70 percent supermajorities of both houses of the California legislature.

The first meeting of the oversight committee had to be cut embarrassingly short when it was discovered that the way it was announced had violated the state's open-meeting rules.[18] And the election of a chair turned farcical when all four officials charged with nominating a candidate happened to pick Robert Klein, who had been the driving force behind the initiative, and had given $2.8 million to that effort, as well as large political donations to at least three of those who nominated him.

Supporters trust that this committee will establish fair and ethical guidelines for disbursing the money and overseeing the research. Opponents remained skeptical.

▪ FUNDING AS A FORM OF REGULATION ▪

PRESIDENT CLINTON, in the immediate aftermath of the first cloned sheep in 1997, did issue an executive order preventing the use of federal funds for human cloning. As he implicitly acknowledged, however, this was largely symbolic:[19] "I am urging the entire scientific community—every foundation, every university, every industry that supports work in this area—to heed the federal government's example."

When President Bush put a similar stop to funding of stem cell research (see **Chapter 4**), it had much more impact, because he was going against rather than along with the scientific consensus. Almost everyone likely to work in that field depended to some extent on federal funds; to continue, even with private money, they had to establish new laboratories that were in no way supported by National Institutes of Health (NIH) grants.

Conversely, federal decisions can have enormous, and sometimes unanticipated, ripple effects. In 1979, the government ruled that retirement funds could, for the first time, invest in venture capital businesses.[20] That helped to provoke the first wave of what Brian Alexander has called "biomania," as hundreds of millions of dollars began to look for new homes.

In 1985, another seemingly innocuous decision allowed NIH scientists to consult for private industry. By the end of 2004, the consequences of this were blooming into a full-grown scandal, as it was revealed that researchers had accepted up to $500,000 in fees from the very companies whose products they were evaluating.[21]

■ PATENTS ■

ANOTHER IMPORTANT ELEMENT in the management of developing technologies is patent law. Two major changes affecting biological patents occurred in 1980: passage of the Bayh-Dole Act, and a Supreme Court ruling that living creatures could be patented.

The Bayh-Dole Act not only enabled but encouraged universities and other research centers to own—and to make money out of—the patents on discoveries they had made using federal grants. The general idea was that the public would benefit from technology being turned into products. In the case of biology, Stuart Newman has noted that this[22] "impressed a commercial stamp on much of the new biology. In particular, informal scientific give-and-take that had characterized biological research in earlier periods was curtailed and conflict of interest concerns that were previously unknown to fields such as cell and development biology became prominent."

It is not clear that the public as a whole has benefited. Indeed, some critics have suggested that patents actually slow down innovation.[23] Others have suggested that there might be more worthwhile motives for discovering, say, a cure for cancer than just the patent rights.

Within limits, however, that is the way the system works. But what are the limits? Should corporations (or individuals) be able to patent life? Traditionally, the Patent Office had always held that living creatures could not be patented. That changed in 1980, when the Supreme Court allowed a patent for a bacterium that had been genetically engineered to be better at eating crude oil, specifically to clean up oil spills (*Diamond v. Chakrabarty*).[24]

In 1988, the first patent for a mammal was issued—for the OncoMouse, a strain of GE mice that develop cancer easily and thus are useful for research. But if you want to use the Onco-Mouse for research on fighting cancer, that means you have to deal with the pharmaceutical giant DuPont, which has an exclusive deal with Harvard (where it was developed) to license the mouse. Other researchers have complained that DuPont "has become more aggressive, pressuring academics to sign restrictive license agreements even though the scientists believe their work falls outside of the OncoMouse patents."[25]

The patent is still controversial. The Canadian Supreme Court rejected it in 2002, after a very long court battle; a patent was issued in Europe, but what it covers may be less than what is covered in the US. (The full decision, with its reasoning and implications, had not been published at this writing.)

Patent law in this area is still very much in flux. Stem cell lines, for example, are likely to provoke what lawyers regard as "interesting" questions, especially if their use does lead to commercially valuable therapies.

■ THE NIH, RAC, CDC, AND FDA ■

IN THE EARLY 1970S, as the technology of genetic engineering was being developed, there was considerable concern, among the public and to some extent among the scientists performing the research. As Paul Berg, who won his Nobel Prize for work on the biochemistry of DNA, later wrote,[26] "Fears of creating new kinds of plagues or of altering human evolution or of irreversibly altering the environment were only some of the concerns that were rampant."

Berg and others called in the journals *Science* and *Nature* for a voluntary moratorium on the work to consider these issues. The NIH stepped in and established, on October 7, 1974, the Recombinant DNA Advisory Committee (RAC). The following February, the NIH sponsored the Asilomar Conference, at which leading scientists discussed the issues.

The scientists' initative was real, and so was the government response, which in the long run may have been more important. The RAC, over the next several years, developed guidelines for genetic research, and still oversees "all human gene transfer trials in which NIH funding is involved (either directly or indirectly)."[27] Note—once again—the extent to which regulation is dependent on funding sources.

A substantial part of the RAC's focus is to consider the ethical implications of any new gene-transfer technique. The results of its deliberations go to the NIH, which establishes guidelines, notably on human experiments (see **Box 12.2**).

EXPERIMENTS ON HUMANS

THERE ARE FEDERAL GUIDELINES for experiments involving human subjects, codified by the National Institutes of Health (NIH). They are based on several foundational documents, especially the Nuremberg Code, the Declaration of Helsinki, and the Belmont Report.

Medical ethics has a history going back much further, at least to the Hippocratic Oath, which dates to about 400 B.C. There was not, however, very much concern about research medicine until the aftermath of World War II, when it became known that Nazi doctors had horribly abused prisoners in what they claimed were scientific experiments.

The Nazis were tried at Nuremberg, and during the trial Directives for Human Experimentation, called the **Nuremberg Code**, were formulated to give basic ethical standards for experimenting on humans. There are ten simple rules, of which the first and most important begins:[28]

1. The voluntary consent of the human subject is absolutely essential.

The Code stresses at some length that "voluntary" means with full knowledge and without any kind of "force, fraud, deceit, duress, over-reaching, or other ulterior form of constraint or coercion." It also insists that experiments be necessary, useful, properly prepared, and conducted by qualified personnel so as to avoid all possible suffering.

The Nuremberg Code was elaborated in the 1964 **Declaration of Helsinki**, by the World Medical Association. This stressed that "the importance of the objective [must be] in proportion to the inherent risk to the subject," and that the subject must be fully informed and likely to benefit. It concludes, "In research on man, the interest of science and society should never take precedence over considerations related to the well-being of the subject."

Prompted in part by the scandal of the Tuskegee syphilis studies (see **Chapter 11**), the US government formed the National Commission for the Protection of Human Subjects of Biomedical and Behavioral Research, which met from 1974 to 1978 and produced an authoritative document known as **The Belmont Report**: Ethical Principles and Guidelines for the Protection of Human Subjects of Research. This established three basic ethical principles:

- **respect for persons**—the rights of research subjects, especially those with diminished autonomy and capacity
- **beneficence**—research must not only avoid harming those involved but must also be intended to help
- **justice**—just distribution of potential benefits and harms and fair selection of research subjects

As summarized by the President's Council on Bioethics,[29] "When applied, these general principles lead to both a requirement for informed consent of human research subjects and a requirement for a careful assessment of risks and benefits before proceeding with research. Safety, consent, and the rights of research subjects are thus given the highest priority."

12.2 The principles of the Belmont Report are not universally accepted. They have been criticized for a utilitarian bias and for their practical effect of limiting the discourse of bioethics. Nevertheless, they have become, as John Evans notes, the "public law governing the research activities of federally funded scientists . . . the standard not only for federally funded research, but for privately sponsored research as well."[30]

Safety and effectiveness are fundamentally the purview of the Food and Drug Administration (FDA). If any gene-therapy products were to be approved for use with humans (as opposed to the experiments that have been taking place since 1990; see **Chapter 6**), it would be the FDA that certified them.

The Centers for Disease Control and Prevention (CDC), which is part of the US Department of Health and Human Services, also has a role, and a potentially significant one: It collects data on the success rates of fertility clinics and publishes them on its website, www.cdc.gov/reproductivehealth/art.htm. This is useful, but hampered by the fact that some clinics get away with not reporting. For example, the Kentucky Center for Reproductive Medicine is the clinic run by Dr. Zavos, who wants to clone people; it had not been included in the Reports published through 2004. As oversight, it's minimal.

These systems may have been adequate in the 1970s. They seem—as the cloning farce described in **Box 12.1** shows—sadly short of what is required in the twenty-first century.

■ PROFESSIONAL SELF-REGULATION ■

MEDICAL PRACTITIONERS HAVE a complex system of self-regulation, and fertility doctors are no exception. The limitations of this were graphically demonstrated in late 2001, when—with no apparent warning—the American Society for Reproductive Medicine (ASRM) announced that it was at least sometimes acceptable to perform sex selection.[31]

That provoked a storm of protest and a rapid repudiation by the ASRM's Executive Director, who claimed that the previous comments had been "taken out of context."[32] Nevertheless, the practitioner whose enquiry prompted the initial response could still offer sex selection—and in fact does.

Meanwhile, as the legal status of embryonic stem cell research remained vague on the federal level, the National Academy of Sciences has been reported to be developing voluntary guidelines.[33] This is rather unlikely to reassure the public at large, however, which "believes that scientists do not have internally or externally imposed ethical limits on their research, and so they cannot be trusted," according to a 2003 survey by the Genetics and Public Policy Center, whose Director, Kathy Hudson, commented,[34] "The fact that the American public is so distrustful of the scientific community, I find as a geneticist, really disturbing."

That is precisely why scientists could benefit from transparent regulation, publicly discussed and decided in an open political process.

■ SPECIFIC LAWS IN OTHER COUNTRIES ■

AT LEAST FORTY-NINE countries—mostly in Europe—have passed laws to address the issues of reproductive cloning, research cloning or human germline intervention. The details vary, but the Institute on Biotechnology and the Human Future summarizes the position thus,[35] "There is a clear trend in the international context toward a prohibition on all use of human cloning."

National views on reproductive cloning are essentially unanimous. Of forty-nine countries surveyed in 2004:[36]

▶ Forty-one had explicitly banned reproductive cloning by law
▶ One more had placed a moratorium on it
▶ Seven had prohibited it implicitly or under guidelines
▶ none explicitly allowed reproductive cloning

In addition, at the end of 2004, several more countries had legislation in process, or recently enacted, to ban reproductive cloning. Others had ratified the Council of Europe's *Convention* banning cloning and germline intervention (see below) without enacting specific legislation. Absolutely none had made it legal.

Cloning for research purposes is somewhat more contentious. Of the same group of forty-nine countries:

- at least eighteen had explicitly banned research cloning by law
- at least four had prohibited it implicitly or under guidelines
- at least sixteen did not have legislation in place
- at least six allow research cloning

Austria, Costa Rica, Slovakia, and South Africa had prohibited research cloning implicitly or under guidelines. The following countries had banned it explicitly:

Argentina	Germany	Norway
Canada	Iceland	Peru
Denmark	Italy	Romania
Estonia	Japan	Slovenia
Finland	Lithuania	Spain
France	Netherlands	Switzerland

The five that explicitly allow research cloning (generally under restrictions) were:

Belgium	Cuba	UK
China	Singapore	South Korea

Israel, Sweden, and perhaps others may also permit research cloning. There is pressure to allow it, even in Germany, which has a strong law against it. According to a spokesman for the German Federal Reserach Ministry, however, "There is no majority for changing the ban on cloning."[37]

On germline engineering, those—relatively few—countries that have enacted legislation are unanimous. Of the same forty-nine:

▶ Twenty-one had banned germline engineering
▶ Two more had prohibited it implicitly or under guidelines
▶ Twenty-five did not have legislation in place
▶ None had explicitly allowed germline engineering

A majority of nations, especially in less developed countries, have not yet enacted legislation on these topics, although there are discussions at the African Union.[38] In Europe, however, most countries have. There are also international agreements within Europe.

■ THE COUNCIL OF EUROPE ■

THE OLDEST AND LARGEST European political organization is the Council of Europe (CoE), which has forty-six members, as of the start of 2005. Belarus has also applied to join, and is the only European country that is not yet a member, unless you count Kazakhstan, which is mostly in Asia, and the Vatican. The Vatican does have observer status, as do the US, Canada, Mexico, and Japan. The CoE is much larger than the European Union, and focuses primarily on human rights and the promotion of democracy.

One of the first major actions of the CoE, in 1950, was the *Convention for the Protection of Human Rights and Fundamental Freedoms*. In the 1990s, members negotiated a *European Convention on Human Rights and Biomedicine*, and after the announcement of the first cloned sheep rapidly added an *Additional Protocol*, which prohibits the cloning of human beings and firmly places this ban in the context of human rights. Taken together, these have been described as "the most comprehensive multilateral treaty addressing the new human genetic technologies."[39]

The *Human Rights and Biomedicine Convention* bans genetic discrimination and the abuse of genetic testing and biomedical experiments, and includes:[40]

> An intervention seeking to modify the human genome may only be undertaken for preventive, diagnostic or therapeutic purposes and only if its aim is not to introduce any modification in the genome of any descendants.
>
> The use of techniques of medically assisted procreation shall not be allowed for the purpose of choosing a future child's sex, except where serious hereditary sex-related disease is to be avoided. . . .
>
> The creation of human embryos for research purposes is prohibited.

The *Convention* had been signed by thirty-one of the Council's forty-six member states, and ratified by nineteen, as of January 1, 2005: Bulgaria, Croatia, Cyprus, the Czech Republic, Denmark, Estonia, Georgia, Greece, Hungary, Iceland, Lithuania, Moldova, Portugal, Romania, San Marino, Slovakia, Slovenia, Spain, and Turkey. The *Additional Protocol* had been ratified by fifteen of its twenty-nine signatories (the same list as has ratified the *Convention*, less Bulgaria, Denmark, San Marino, and Turkey).

The list of signatories is, however, notable for omissions. Some nations, such as Germany, have passed stronger laws, and chose not to sign the *Convention* for fear of weakening their position; others, such as the UK, took the view that some of the provisions went too far, notably the ban on creating embryos for research purposes. That might not prevent research on "leftover" embryos that were created in the course of fertility treatment, but it is broader than a ban on cloning embryos. Other states that have not ratified the *Convention* include Belgium, France, Ireland, Italy, the Netherlands—all members of the European Union, where these topics remain controversial.

■ THE EUROPEAN UNION ■

THE EUROPEAN UNION (EU) has twenty-five member states, including most of the major European nations, with a total 2004 population of 456 million. There is a complex system of government, which includes a European Parliament, directly elected by the citizens of the member countries. This has some authority over member states, but in many matters they retain the power to make their own laws.

Several times between 1997 and 2000, the European Parliament called on the nations of the EU "to enact binding legislation prohibiting all research into any kind of human cloning within its territory and providing for criminal penalties for any breach."[41] As part of its 1998 Directive on Patents, the Parliament also declared:[42] "There is a consensus within the community that intervention in the human germ line and the cloning of human beings offends against the *ordre public* and morality." ("Ordre public" is untranslatable—it does mean "public order" but roughly in the sense of "community standards," not the opposite of "riot.") The claimed consensus does not, however, stretch to the issue of embryonic stem cell research, let alone cloning for research purposes. In 2003, the European Parliament once again voted in favor of a ban on the creation of human embryos for research purposes. It did, however, vote to allow states to fund embryonic research that does not involve cloning.[43]

▪ UNESCO ▪

THE UNITED NATIONS Educational, Social, and Cultural Organization (UNESCO) developed a *Universal Declaration on the Human Genome and Human Rights* in 1997. This bases its considerations of these issues very firmly in social justice grounds, with an emphasis on universal rights. Article 1 declares:[44] "The human genome underlies the fundamental unity of all members of the human family, as well as the recognition of their inherent dignity and diversity. In a symbolic sense, it is the heritage of humanity."

The *Declaration* condemns the commercialization of the human genome, genetic discrimination, research without consent, and other abuses. It calls for particular care in the conduct of scientific research and specifically states in Article 11: "Practices which are contrary to human dignity, such as reproductive cloning of human beings, shall not be permitted."

It also instructs UNESCO's Bioethics Committee to investigate further and advise "regarding the identification of practices that could be contrary to human dignity, such as germ-line interventions" (Article 24).

▪ THE WHO ▪

THE WORLD HEALTH ORGANIZATION (WHO)—technically, its deliberative body, the World Health Assembly—also condemned reproductive cloning in 1997. The Resolution stated that,[45] "the use of cloning for the replication of human individuals is ethically unacceptable and contrary to human integrity and morality."

A 2002 WHO report, *Genomics and World Health*, reiterated the view that:[46] "reproductive cloning would be unethical under any circumstances and that there is no ethical or medical basis for pursuing work on it." It did, however, imply that others considered the only problem with cloning "at the present time" to be the risks involved, and acknowledged without particular criticism that, "some would not even rule out the possibility of germ-line enhancement in the future."

The report even began to lay down some guidelines for the application of possible future enhancement technologies, should they become feasible: "If at some point in the future it becomes possible safely to enhance, for example, individuals' memory, intelligence, or immune system, doing so would likely be beneficial to almost everyone in most social contexts. These traits are 'all purpose means,' useful in nearly any plan of life. Any enhancements of children undertaken by their parents should be of traits like these." These speculations flatly contradict the Council of Europe *Convention* and the sense of the UNESCO *Declaration*. The WHO report also, rather hopefully, asserted that, "the distinction between [reproductive and research cloning] is absolutely clear and would provide no problems for appropriate legislation."

That this statement is, in practice, overly optimistic is shown by the experience of not only the US and the EU but also the UN itself.

■ THE UN ■

THE UNITED NATIONS (UN) General Assembly set up an ad hoc committee in 2001 "for the purpose of considering the elaboration of an international convention against the reproductive cloning of human beings."[47]

France and Germany proposed this initiative, which was originally intended to ban reproductive cloning and to establish a subsequent process to consider research cloning. Both France and Germany ban research cloning—Germany has what may be the strictest law in the world—but they considered that it was worthwhile to postpone dealing with the more contentious issue in order to get a ban on reproductive cloning passed.

In short, the US scuppered that deal. President Bush's policy opposed both research cloning and reproductive cloning, and US representatives began lobbying hard for a ban on both; they got enough support—initially about forty countries—to block consensus.[48] That delayed negotiations, and various attempts were made to produce a compromise. They failed. In November 2003, by one vote (80–79, with 15 abstentions and 17 not present), the committee decided to postpone consideration until 2005.

In March 2005, the General Assembly finally passed a nonbinding resolution calling for a ban on "all forms of human cloning in as much as they are incompatible with human dignity" by 84–34, with 37 abstentions.[49] The Bush administration claimed victory, but advocates of research cloning, including Britain, China, and Belgium, immediately said they would ignore it.

President Bush's stance was widely resented in other countries. The Singapore representative, for example, complained about "an all-or-nothing attitude [that] paralyzed the process," while the Turkish spokesman for the Islamic Conference complained about the "negative atmosphere" in such a polarizing setting.[50]

Others were less restrained. In the London *Times*, a column on the November 2004, UN vote was headlined,[51] "Bring On the Clones: The World Decides to Ignore George Bush"

The world does, however, have better examples to offer of a process to get sensible regulation of human genetic technologies. Two are particularly useful for Americans to consider: the systems in place in the UK and Canada.

■ THE UK HFEA ■

IN 1982, the British government set up a Committee of Inquiry into Human Fertilisation and Embryology, chaired by Baroness Mary Warnock, "to consider recent and potential developments in medicine and science related to human fertilisation and embryology; to consider what policies and safeguards should be applied, including consideration of the social, ethical and legal implications of these developments; and to make recommendations."[52]

In 1984, the Warnock Report, as it is normally known, made several controversial recommendations:

- to ban commercial surrogacy
- to allow research on embryos up to fourteen days after fertilization
- to establish a licensing authority to regulate reproductive technologies

The surrogacy ban was enacted in 1985; the fourteen-day limit is still in force, and still offensive to some "pro-life" campaigners; and the UK Human Fertilisation and Embryology Authority (HFEA) was finally established in 1991, following a decade of debate. The HFEA's principal tasks are to:[53]

- License and monitor clinics that carry out in vitro fertilization (IVF) and donor insemination
- License and monitor research centers undertaking human embryo research
- Regulate the storage of gametes and embryos

It is also charged with various informational services, both to the public and to the government. It aims "to safeguard the interests of patients, children, the general public, doctors, service providers, the scientific community, and also future generations."

The HFEA frequently conducts public consultations, on subjects including sex selection and the donation of sperm and eggs. It licenses—and limits—the use of pre-implantation genetic diagnosis, requiring a minimum of two peer reviewers and a committee to consider "the scientific, ethical and medical information" for any application. In August 2004, it controversially issued the first license for cloning to produce embryonic stem cells for research purposes. It has been criticized by some scientists for being too bureaucratic and cumbersome, and by some campaigners for being both secretive and too permissive.[54]

Certainly the British system has demonstrated that government regulation need not inhibit research. At least one important stem cell researcher, Dr. Roger Pedersen, moved from the University of California at San Francisco to become Professor of Regenerative Medicine at Cambridge University, specifically because of the regulatory uncertainty in the US.[55]

The House of Commons Science and Technology Committee is, at this writing in early 2005, completing a major review of the HFEA's role, which is expected to evolve. The British government has proposed a Human Tissue Authority (HTA) and has discussed plans to merge the HFEA with the HTA to form the Regulatory Authority for Fertility and Tissue (RAFT). Nevertheless, the HFEA is frequently held up as a model for other countries, including the US. Certainly Canada took note of the British experience when, in 2004, it finally set up its own system.

■ THE CANADIAN AHRAC ■

CANADA SPENT even more time establishing a regulatory system for human reproductive technologies than the UK did—at least fifteen years. The Assisted Human Reproduction Act (AHRA) was finally passed in March, 2004; it includes a mandated review after three years, so in a sense the process is still not complete.

The Canadian Prime Minister set up a Royal Commission on the new reproductive technologies at the end of 1989.[56] This was an independent body, with a $30 million budget, which actively sought public input, through hearings, toll-free hotlines, public surveys, and research programs, as well as written briefs;

an estimated 40,000 people contributed. It reported at the end of 1993, but no comprehensive bill was introduced until 1996.

That bill was opposed by, among others, the Commission's head because it focused too much on prohibitions and not enough on ensuring the safe use of benevolent technologies. It died with the Parliamentary elections of 1997, but public pressure for a resolution continued. The final 2004 version contains a balance of prohibitions and regulations, to be overseen by the Assisted Human Reproduction Agency of Canada (AHRAC). The stated goals are:[57]

▶ to protect the health and safety of Canadians;
▶ to prevent commercial exploitation of reproduction; and
▶ to protect human individuality and diversity and the integrity of the human genome

The AHRAC will issue licenses to all fertility clinics and oversee all embryo research, which is legal but limited. It will also set standards for the use of pre-implantation genetic diagnosis (PGD) and other such technologies, including limits on the number of embryos that can be implanted at one time. It will maintain records and, crucially, require fully informed consent from all patients and donors.

Among the prohibited practices are:[58]

▶ the creation of human embryos just for research
▶ cloning, whether for reproduction or for research
▶ the creation of human/non-human hybrids and chimeras
▶ inheritable genetic modification (germline engineering)
▶ sex selection except in the case of sex-linked disease
▶ commercial surrogate motherhood contracts
▶ the sale of sperm, eggs, and embryos

The bans have teeth—jail terms of ten years and fines of half a million dollars for contravening the major provisions. The primary focus of the legislation is, however, positive: It is intended to encourage safe, ethical use of the new technologies. Reasonable compensation for egg donation is allowed, for example; outright sale, with the attendant risk of exploitation, is forbidden. Research on embryos originally created in fertility treatment is allowed; making them specially is forbidden.

The details are still being worked out, in a further round of public consultation. This process, and the three-year review, helped to solidify the support of several interest groups that do have criticisms of the law, some of which are mutually contradictory. Some scientists object to the criminal penalties; some anti-abortion campaigners to all embryo research; some disability activists consider the rules about PGD to be too lax; some infertile couples want to be able to pay for women's eggs. But overall, a consensus seems to have been achieved that, as Abby Lippman and Diane Allen—Chair of the Canadian Women's Health Network and Executive Director of The Infertility Network, respectively—wrote as the Bill was being considered in the Canadian Parliament,[59] "Canadians have been living in a legal and ethical vacuum with respect to assisted human reproduction for far too long."

Can the US find the political wisdom to produce its own solution?

■ FURTHER READING ■

Free Documents from the Web

The invaluable Global Lawyers and Physicians for Human Rights Database of Global Policies on Human Cloning and Germline Engineering is at www.glphr.org/genetic/genetic.htm.

The President's Council on Bioethics, *Reproduction and Responsibility: The Regulation of New Biotechnologies*, is a 305-page report, available as a single pdf (2.2mb) or a series of web pages from www.bioethics.gov/reports/reproductionandresponsibility/index.html.

M. Asif Ismail, "In Congress, a Cloning Stalemate," Center for Public Integrity, 06/02/04, one of a six-part series, all available from www.publicintegrity.org/genetics/report.aspx?aid=281&sid=200

David Appell, "The New Uncertainty Principle," *Scientific American*, 01/01; available at www.mindfully.org/Precaution/Uncertainty-Principle.htm

Mark Dowie, "God and Monsters," *Mother Jones*, 01–02/04; motherjones .com/news/feature/2004/01/12_401.html. This entertaining article describes the efforts of Stuart Newman and Jeremy Rifkin to patent a chimera, an animal-human hybrid, specifically in order to ban its production. They later declared victory and quit, when the Patent Office rejected the application.

Books

Andrew Kimbrell, *The Human Body Shop*, Regnery Publishing, 1997

William Kristol and Eric Cohen, eds., *The Future Is Now: America Confronts the New Genetics*, Rowman & Littlefield, 2002

Sheldon Krimsky, *Science in the Private Interest: Has the Lure of Profits Corrupted Biomedical Research?* Rowman & Littlefield, 2003

■ ENDNOTES ■

1 Lori B. Andrews, J.D., and Nanette Elster, J.D., M.P.H., "Regulating Reproductive Technologies," *The Journal of Legal Medicine*, 21:35–65, 2000

2 genetics-and-society.org/policies/other/canada.html.

3 *Los Angeles Times*, 03/29/01

4 Charleston Sunday *Gazette-Mail*, 08/05/01

5 *U.S. News & World Report*, 07/09/01

6 *Washington Post*, 03/29/01

7 *Boston Globe*, 04/04/01

8 *Syracuse Post-Standard,* 07/15/01.

9 M. Asif Ismail, "In Congress, a Cloning Stalemate," Center for Public Integrity, 06/02/04; the bills are listed at www.publicintegrity.org.

10 Tom Abate, "U.S. Moratorium on Cloning Sought," *San Francisco Chronicle*, 03/20/02.

11 Mary Leonard, "Coalition Urges a Ban on All Human Cloning," *Boston Globe*, 03/22/02

12 Based on data collected by the National Conference of State Legislatures, available at ncsl.org/programs/health/genetics/rt-shcl.htm

13 Martin Kasindorf, "States Play Catch-Up on Stem Cells," *USA Today*, 12/16/04

14 M. Asif Ismail, "The Biotech Industry Pushes Its Agenda in the States," Center for Public Integrity, 03/02/04

15 See summary and extensive links at genetics-and-society.org/policies /california/index.html

16 California Secretary of State Campaign Contributions Database, accessed 03/03/05; http://dbsearch.ss.ca.gov/BallotSearch.aspx

17 Tali Woodward and Laura M. Allen, "Second-Guessing Prop. 71," *San Francisco Bay Guardian*, 12/22/04

18 Carl T. Hall, "Prop. 71 Financier Debuts as Chairman," *San Francisco Chronicle*, 12/18/04

19 Rick Weiss, "Clinton Forbids Funding of Human Clone Studies," *Washington Post*, 03/05/97

20 Brian Alexander, *Rapture: How Biotech Became the New Religion*, Basic Books, New York, 2003, p. 81

21 David Willman, "The National Institutes of Health: Public Servant or Private Marketer?" *Los Angeles Times*, 12/22/04

22 Stuart A. Newman, "Averting the Clone Age: Prospects and Perils of Human Developmental Manipulation," *Journal of Contemporary Health Law and Policy*, Spring 2003, pp. 431–463

23 Michael A. Heller and Rebecca S. Eisenberg, "Can Patents Deter Innovation? The Anticommons in Biomedical Research," *Science*, 05/01/98; the question of motives has been raised by Barbara Katz Rothman, among others.

24 Andrew Kimbrell, *The Human Body Shop*, Regnery Publishing, Inc., Washington, DC, 1997

25 Gareth Cook, "OncoMouse Breeds Controversy," *San Francisco Chronicle*, 06/03/02

26 Paul Berg, "Asilomar and Recombinant DNA," an essay at the Nobel Prize website, 08/26/04

27 NIH web-based fact sheet, http://www4.od.nih.gov/oba/rac/about rdagt.htm

28 National Institutes of Health (NIH), *Guidelines for the Conduct of Research Involving Human Subjects*, 5th printing, 08/04; available as a pdf from http://ohsr.od.nih.gov/guidelines/guidelines.html, where can also be found links to the Belmont Report and other documents.

29 President's Council on Bioethics, *Human Cloning and Human Dignity: An Ethical Inquiry*, 88

30 John H. Evans, *Playing God?* University of Chicago Press, 2002, p. 89

31 Gina Kolata, "Fertility Ethics Authority Approves Sex Selection," *New York Times*, 09/28/01

32 "Embryonic Sex Selection: ASRM Does Not Endorse IVF Use," Reuters, 10/02/01

33 "National Academy of Sciences to Develop Voluntary Guidelines for Human Embryonic Stem Cell Research," *Kaiser Daily Reproductive Health Report*, 06/29/04

34 John Schieszer, "Most Americans Oppose Idea of 'Designer Babies,'" Reuters Health, 10/15/03

35 "International Legal Situation—Cloning," Institute on Biotechnology and the Human Future, n.d. (apparently June, 2004); thehuman future.org/topics/humancloning/international.htm

36 This summary is largely based on one made by Rosario Isasi, JD, MPH, which is available, as a pdf, from thehumanfuture.org/topics/human cloning/international.htm. I added Cuba to the list of nations, and checked the data on 12/24/04 against another useful summary—for which Isasi was also lead researcher, along with Jesse Reynolds— available at genetics-and-society.org/policies/survey.html.

37 "Germany's Ethics Council Rejects Cloning," *Deutsche Welle*, 09/13/04. An AP report (02/19/05) estimated that "at least 20 nations . . . favor therapeutic cloning." The legal position continues to be very hard to define.

38 See the Agenda for the December, 2004, meeting of the African Union, at africa-union.org

39 Center for Genetics and Society policy summary; genetics-and-society.org/policies/international/council.html

40 "Convention for the protection of Human Rights and dignity of the human being with regard to the application of biology and medicine: Convention on Human Rights and Biomedicine," summary, text, lists of signatories, etc, linked from http://conventions.coe.int/Treaty/Commun/QueVoulezVous.asp?NT=164&CM=8&DF=25/12/04&CL=ENG

41 Rory Watson, "EU Institutions Divided on Therapeutic Cloning," *British Medical Journal*, 09/16/00

42 European Parliament Patent Directive 98/44/EC, 07/06/98

43 "Europe Backs Embryonic Stem Cell Research," *New Scientist*, 11/19/03

44 United Nations Educational, Social, and Cultural Organization (UNESCO), *Universal Declaration on the Human Genome and Human Rights*, 11/11/97; available from www.unesco.org/shs/bioethics

45 "World Health Assembly States Its Position on Cloning in Human Reproduction," Press Release, 05/14/97; www.who.int/archives/inf-pr -1997/en/97wha9.html

46 World Heath Organization Advisory Committee on Health Research, *Genomics and World Health*, Geneva, 2002; available as a single large pdf or 14 smaller ones from http://www3.who.int/whosis/genomics/genomics_report.cfm. The quotations are all from sections 8.7 and 8.8.

47 United Nations General Assembly, "Summaries of the Work of the Sixth Committee," un.org/law/cod/sixth/59/summary.htm#153

48 Center for Genetics and Society, "2003 Special Report on the UN Cloning Treaty Negotiations," genetics-and-society.org/policies/international/2003unreport.html

49 "US Gets Cloning Ban Victory at UN," AFP, 03/08/05

50 David Dickson, "Cloning Ban Delay Is 'Tremendous Victory,'" South Africa *Mail & Guardian*, 11/10/04

51 Anjana Ahuja, "Bring on the Clones: The World Decides to Ignore George Bush," London *Times*, 11/29/04

52 From a slide presentation by Henry Leese, of the HFEA, made in September 2002 and archived at www.fda.gov/cber/summaries/art091802hl.htm

53 From the HFEA website, www.hfea.gov.uk/AboutHFEA

54 For trenchant criticism from Lord Winston, a fertility expert, see: "Fertility Watchdog 'Incompetent,'" BBC, 12/10/04; http://news.bbc.co.uk/1/hi/health/4084365.stm. Josephine Quintavalle, of Comment on Reproductive Ethics (CORE), commented, "For once, we half-agree with him," in *Medical News*, 12/10/04

55 Tom Abate, "UCSF Stem Cell Expert Leaving U.S.," *San Francisco Chronicle*, 07/17/01

56 For a historical overview, see the testimony of Dr. Patricia Baird, who chaired the Commission, before the US President's Council on Bioethics, 06/20/02, available at bioethics.gov/transcripts/jun02/june20 session1.html

57 ibid.

58 See the summary at genetics-and-society.org/policies/other/canada.html, which includes a comparison with the German and Australian systems, as well as the British.

59 Abby Lippman and Diane Allen, "Time to Fill the Legal and Ethical Vacuum on Stem-Cell Research," *The Hill-Times*, 09/01/03

APPENDIX: RESOURCES

SPECIALIST WEBSITES

The Center for Genetics and Society (CGS) • genetics-and-society.org
a large and well-organized website, with links to hundreds
more; good summaries of the scientific and political issues,
and the responses of various constituencies; email newsletter
listed below

The Council for Responsible Genetics (CRG) • gene-watch.org
publisher of *GeneWatch*, "America's first and only magazine
dedicated to monitoring biotechnology's social, ethical and
environmental consequences," six times a year, subscriptions
from $35; website includes archived issues

President's Council on Bioethics • bioethics.gov
a remarkable resource; several comprehensive reports, available
free, with background papers and transcripts of presentations
made at its meetings

Institute on Biotechnology and the Human Future (IBHF) •
thehumanfuture.org
particularly strong on legal issues; also reposts some important
general papers

EMAIL NEWSLETTERS

Genetics Crossroads • genetics-and-society.org/newsletter
from CGS; six times a year, plus occasional special issues; usually includes one or more features, plus news, often with
detailed analysis

Bioethics News Highlights • cbhd.org/email/index.html
from the Center for Bioethics and Human Dignity (CBHD);
weekly, the first sentence or two of about 15 news articles,
plus links, presented without comment; plus a monthly Update
on the Center's activities, with Christian commentary

Human Genetics Alert (HGA) News Service •
hgalert.org/join/how_to_subscribe.htm
one mail containing the complete text of about 6 articles, plus
brief comments, from a British activist group; used to be daily,
became weekly in 2004

GM Watch • gmwatch.org
daily, weekly or monthly; a terrific resource, focused on GE
Food but also covering Human GE; British-based but with a
very international approach, including US activities and particularly strong on Africa and Asia; up to 5 emails a day, or
single-email weekly or monthly round-ups

Organic Bytes • organicconsumers.org
twice a month from the Organic Consumers Association
(OCA); one email with a paragraph (and picture) per story and
links to the rest; pdf version also available; mostly on food and
agriculture

ACTIVISTS

Many social-justice, environmental, feminist and other organizations have interests in Human GE issues (see **Chapter 11**). As well as such prominent ones as **Friends of the Earth** (foe.org) and **Greenpeace** (greenpeaceusa.org), which have large agendas, these are among the most important, and all run by activists with a long-standing commitment to these issues:

Our Bodies Ourselves (OBOS), aka the Boston Women's Health Book Collective (BWHBC) • ourbodiesourselves.org

International Center for Technology Assessment (ICTA) • icta.org

The Foundation on Economic Trends • foet.org

Institute for Social Ecology (ISE) • social-ecology.org (Biotechnology Project)

Disabled Peoples International (DPI) • dpi.org

Indigenous Peoples Council on Biocolonialism (IPCB) • ipcb.org

RELIGIOUS PERSPECTIVES

The Pew Forum on Religion and Public Life • pewforum.org
useful overviews in the 'Publications' section

The Center for Bioethics and Human Dignity (CBHD) • cbhd.org
a resource particularly for Christian bioethics, with position statements and an excellent series of links as well as an email news service

CHRISTIAN DENOMINATIONS

The United States Conference of Catholic Bishops • usccb.org/prolife/issues/bioethic

United Methodist Church • umc.org (search for 'genetics' or 'cloning')

Baptists for Life, Center for Biblical Bioethics • bfl.org/cbb

Southern Baptist Convention • sbc.net/resolutions (search by the topic 'genes')

United Church of Christ • ucc.org/justice/stemcell

Episcopal Church • episcopalchurch.org (search for 'genetics')

OTHER RELIGIONS

Religious Action Center of Reform Judaism • rac.org/advocacy/issues/issuebe

Jewish Law, preliminary analyses • jlaw.com/Articles/cloning.html jlaw.com/Articles/stemcellres.html

A collection of links to Islamic views • islamicmedicine.org/views.htm#gen

Hindu perspectives • hinduismtoday.com/archives/1997/6/1997-6-11.shtml

US GOVERNMENT AGENCIES

National Institutes of Health (NIH) • nih.gov
This enormous website is rather confusingly ordered, with many of the sub-sites set up so that 'www.' must *not* be used.

Stem Cells at NIH • http://stemcells.nih.gov

Human Genome Project (HGP) • genome.gov; also ornl.gov/sci/techresources/Human_Genome/home.shtml

Centers for Disease Control (CDC), Reproductive Health Information and Reports • cdc.gov/reproductivehealth/art.htm

Food and Drug Administration (FDA) • fda.gov

OTHER WEBSITES

The World Anti-Doping Agency (WADA) • wada-ama.org

The Pew Initiative on Food and Biotechnology • pewagbiotech.org
claims to be impartial but in practice seems to favor biotech; good survey data

Genetics and Public Policy Center (GPPC) • dnapolicy.org
at Johns Hopkins, funded by the Pew Charitable Trusts

The Corner House • thecornerhouse.org.uk/subject/genetics
a radical British think-tank that produces excellent briefing papers

Genetic Engineering and Its Dangers • http://online.sfsu.edu/~rone/GEessays/gedanger.htm
a large collection of links and a list of books, compiled by Professor Ron Epstein

GE FOOD CAMPAIGNERS

Agricultural biotech provoked a number of new activist groups, supplementing efforts by previously established organizations like Greenpeace and Friends of the Earth. Sites worth checking for further information include:

The Center for Food Safety (CFS) • centerforfoodsafety.org

The Organic Consumers Association (OCA) • organicconsumers.org

The Campaign to Label Genetically Engineered Foods • thecampaign.org

The Pesticide Action Network North America (PANNA) • panna.org

The Institute for Food and Development Policy (Food First) • foodfirst.org

The GE Food Alert Coalition • gefoodalert.org

The Genetic Engineering Action Network (GEAN) • geaction.org
a network of about 100 organizations, including many local groups; distributes a CD that includes about 2,000 articles worth having, compiled by Luke Anderson, author of *Genetic Engineering, Food and Our Environment*

BOOKS

CRITICS AND SKEPTICS

Three books with important but different critical perspectives on Human GE, one from an environmentalist (McKibben), one from an academic conservative (Fukuyama), and one from a free-swinging maverick (Smith) are:

Bill McKibben, *Enough: Staying Human in an Engineered Age*, Henry Holt and Company, 2003

Francis Fukuyama, *Our Posthuman Future: Consequences of the Biotechnology Revolution*, Farrar, Straus and Giroux, 2002

Wesley J. Smith, *Consumer's Guide to a Brave New World*, Encounter Books, 2004

Other significant contributions are:

Lori Andrews, *The Clone Age*, Henry Holt, 1999

Andrew Kimbrell, *The Human Body Shop*, Regnery Publishing, 1997

Sheldon Krimsky, *Science in the Private Interest: Has the Lure of Profits Corrupted Biomedical Research?* Rowman & Littlefield, 2003

Jeremy Rifkin, *The Biotech Century*, Tarcher/Putnam, 1998

Brian Tokar, ed., *Redesigning Life: The Worldwide Challenge to Genetic Engineering*, Zed Books, 2001

Casey Walker, ed., *Made Not Born*, Sierra Club Books, 2000

Several other books take a skeptical position on genetic determinism, notably:

Ruth Hubbard and Elijah Wald, *Exploding the Gene Myth*, Beacon Press, 2nd ed. 1999

Richard C. Lewontin, *Biology As Ideology: The Doctrine of DNA*, Harperperennial, 1993

Richard C. Lewontin, *The Triple Helix: Gene, Organism, and Environment*, Harvard University Press, 2000

Classic essays by Leon Kass and C.S. Lewis can be found in:

Leon R. Kass and James Q. Wilson, *The Ethics of Human Cloning*, The AEI Press, 1998

C.S. Lewis, *The Abolition of Man*, first published in Great Britain in 1943 by Oxford University Press, since reprinted many times by various imprints

Books on eugenics and related topics include:

Edwin Black, *War Against the Weak: Eugenics and America's Campaign to Create a Master Race*, Four Walls Eight Windows, New York, 2003

Stephen J. Gould, *The Mismeasure of Man*, W.W. Norton, revised 1996

Daniel Kevles, *In the Name of Eugenics: Genetics and the Uses of Human Heredity*, Harvard University Press, 1985, revised 1995

SUPPORTERS AND ENABLERS
Among the more extreme are:

Lee Silver, *Remaking Eden: How Genetic Engineering and Cloning Will Transform the American Family*, Bard, 1998; originally published as *Remaking Eden: Cloning and Beyond in a Brave New World*, Avon, 1997

Gregory Stock, *Redesigning Humans: Choosing Our Genes, Changing Our Future*, Mariner Books, 2003; originally published as *Redesigning Humans: Our Inevitable Genetic Future*, Houghton Mifflin, 2002

Gregory Stock and John Campbell, eds, *Engineering the Human Germline: An Exploration of the Science and Ethics of Altering the Genes We Pass to Our Children*, Oxford University Press, 2000

James Hughes, *Citizen Cyborg: Why Democratic Societies Must Respond to the Redesigned Human of the Future*, Westview Press, 2004

Somewhat more subtle are books whose pro-biotech bias is implicit, such as:

Brian Alexander, *Rapture: How Biotech Became the New Religion*, Basic Books, New York, 2003

Allen Buchanan, Dan W. Brock, Norman Daniels and Daniel Winkler, *From Chance to Choice: Genetics and Justice*, Cambridge University Press, 2000

Stephen S. Hall, *Merchants of Immortality: Chasing the Dream of Human Life Extension*, Houghton Mifflin Company, New York, 2003

Gina Kolata, *Clone: The Road to Dolly and the Path Ahead*, William Morrow and Company, 1998

Matt Ridley, *Genome*, HarperCollins, 2000

GE FOOD

The best single book about GE food is:

Jeffrey M. Smith, *Seeds of Deception: Exposing Industry and Government Lies About the Genetically Engineered Foods You're Eating*, Yes! Books, 2003; there is an associated website, seedsofdeception.com, and e-newsletter.

Other useful ones include:

Luke Anderson, *Genetic Engineering, Food and Our Environment*, Chelsea Green Publishing Co., 1999

Ronnie Cummins and Ben Lilliston, *Genetically Engineered Foods: A Self-Defense Guide for Consumers*, Marlowe & Company, 2000

ACKNOWLEDGMENTS

F RIENDSHIP GOT ME into this, which seems entirely appropriate for such a human activity. My friend Alexander Gaguine introduced me to the issue; his support has been vital and I thank him most sincerely. He also introduced me to Richard Hayes and Marcy Darnovsky; their work, along with that of Jesse Reynolds, Sujatha Jesudason, and everyone else at what is now the Center for Genetics and Society, has been a continuing source of inspiration.

Stuart Newman and Ignacio Chapela each offered welcome encouragement right when I needed it. I am grateful to countless other activists, writers, friends, and acquaintances—too many to name except for the most deserving Kaki Rusmore.

At Nation Books, Ruth Baldwin has done an impeccable job of guiding me through the editing and production process, and I thank her as well as Carl Bromley, not to mention John Oakes, who steered me their way.

INDEX